Studies in Computational Intelligence

Volume 711

Series editor

Janusz Kacprzyk, Polish Academy of Sciences, Warsaw, Poland
e-mail: kacprzyk@ibspan.waw.pl

About this Series

The series "Studies in Computational Intelligence" (SCI) publishes new developments and advances in the various areas of computational intelligence—quickly and with a high quality. The intent is to cover the theory, applications, and design methods of computational intelligence, as embedded in the fields of engineering, computer science, physics and life sciences, as well as the methodologies behind them. The series contains monographs, lecture notes and edited volumes in computational intelligence spanning the areas of neural networks, connectionist systems, genetic algorithms, evolutionary computation, artificial intelligence, cellular automata, self-organizing systems, soft computing, fuzzy systems, and hybrid intelligent systems. Of particular value to both the contributors and the readership are the short publication timeframe and the worldwide distribution, which enable both wide and rapid dissemination of research output.

More information about this series at http://www.springer.com/series/7092

Brijesh Verma · Ligang Zhang
David Stockwell

Roadside Video Data Analysis: Deep Learning

Springer

Brijesh Verma
School of Engineering and Technology
Central Queensland University
Brisbane, QLD
Australia

David Stockwell
School of Engineering and Technology
Central Queensland University
Brisbane, QLD
Australia

Ligang Zhang
School of Engineering and Technology
Central Queensland University
Brisbane, QLD
Australia

ISSN 1860-949X ISSN 1860-9503 (electronic)
Studies in Computational Intelligence
ISBN 978-981-10-4538-7 ISBN 978-981-10-4539-4 (eBook)
DOI 10.1007/978-981-10-4539-4

Library of Congress Control Number: 2017936905

Printed on acid-free paper

This Springer imprint is published by Springer Nature
The registered company is Springer Nature Singapore Pte Ltd.
The registered company address is: 152 Beach Road, #21-01/04 Gateway East, Singapore 189721, Singapore

Preface

Video data analysis has become an increasingly important research area with widespread applications in automatic surveillance of transportation infrastructure including roads, rail and airports. As the amount of video data collected grows, so does the opportunity for further processing with new artificial intelligence methods. One type of useful video data that has seen little research is the roadside video data that is collected using video-mounted vehicles. These video data may augment or replace road-based surveys of the conditions of roadside objects such as trees, grasses, roads and traffic signs and can be potentially used in many real-world applications such as roadside vegetation growth condition monitoring, effective roadside management to reduce the possible hazards to drivers and vehicles, and developing automatic vehicles that are able to automatically sense roadside objects and traffic signs.

Most existing studies on video data analysis are primarily focused on analyzing generic object categories in the data content in public benchmark datasets. Very limited research has focused on the analysis of roadside video data, although the significance of developing smart techniques for roadside video data analysis has been widely recognized. One of the main reasons is probably because there is a lack of a comprehensive public dataset that was specifically created for roadside objects. Another reason is the various types of variations and environmental conditions encountered along road sides, which are still challenging issues in the computer vision field. The great variability in the appearance and structure of objects as well as the various types of environmental effects such as underexposure, overexposure, shadows, and sunlight reflectance make accurate segmentation and recognition of objects difficult. The current literature lacks a comprehensive review of existing machine learning algorithms, particularly deep learning techniques, on roadside data analysis.

This book highlights the methods and applications for roadside video data analysis. It describes various system architectures and methodologies that are specifically built upon different types of learning algorithms for roadside video data processing, with detailed analysis of the segmentation, feature extraction and classification. The use of deep learning to solve the roadside video data segmentation and classification problems is one of the major highlights of this book. Deep neural net learning has become popular in machine learning and data mining areas. However, the benefits of a deep feature free approach must be balanced against the considerations of accuracy and robustness, and most real-world learning systems require some hand engineering of features and architectures. This book examines via empirical testing the types of features and architectures that contribute to the performance of multi-layer neural nets on real-world scene analysis. We then demonstrate novel architectures that perform scene classification with equal or better accuracy to previous methods and investigate the feature engineering into convolutional neural networks. Further, we provide an industrial perspective to help align theoretical concerns with real-world results. Finally, as a case study of roadside video data analysis, we demonstrate an application of vegetation biomass estimation techniques for roadside fire risk assessment. Overall, this book compiles the most useful strategies in the field of scene analysis to help researchers identify the most appropriate features and architectures for their applications.

Brisbane, Australia

Brijesh Verma
Ligang Zhang
David Stockwell

Acknowledgements

The authors express their gratitude for help from the Department of Transport and Main Roads in Queensland, Australia for the creation of the data and assistance with field resources for the fire risk survey. This work was supported under Australian Research Council's Linkage Projects funding scheme (project number LP140100939).

Many students and research fellows in the Centre for Intelligent Systems at Central Queensland University have contributed to research presented in this book. The authors would like to thank the following researchers: Dr Sujan Chowdhury, Dr Tejy Kinattukara Jobachan, Dr Peter Mc Leod, Dr M. Asafuddoula, Mrs. Toshi Sinha and Mrs. Fatma Shaheen.

Contents

Abbreviations

ANN	Artificial Neural Network
ANOVA	ANalysis Of VAriance
BG	Brown Grass
CAV	Context Adaptive Voting
CHM	Contextual Hierarchical Model
CIVE	Color Index of Vegetation Extraction
CNN	Convolutional Neural Network
CRF	Conditional Random Field
CRR	Correct Recognition Rate
CSPM	Contextual Superpixel Probability Map
CWT	Continuous Wavelet Transform
DTM	Digital Terrain Model
DTMR	Department of Transport and Main Roads
ExG	Excess Green
ExR	Excess Red
FC	Fully Connected layer
FCM	Fuzzy C-Means algorithm
FRR	False Recognition Rate
GG	Green Grass
GLCM	Gray-Level Co-occurrence Matrix
HOG	Histograms of Oriented Gradients
KNN	K-Nearest Neighbor
LBP	Local Binary Pattern
LIDAR	LIght Detection And Ranging
MLP	Multi-Layer Perceptron
MNDVI	Modification of NDVI
MNIST	Mixed National Institute of Standards and Technology dataset
MO	Morphological Opening
MRF	Markov Random Field
MSRC	MicroSoft Research Cambridge 21-class dataset

N/A	Not Available
NDVI	Normalized Difference Vegetation Index
NIR	Near Infrared Ray
OCP	Object Co-occurence Prior
PHOG	Pyramid Histogram of Oriented Gradient
PID	Pixel Intensity Difference
PPM	Pixel Probability Map
PPS	Pixel and Patch Selective
RBF	Radial Basis Function
RELU	REctified Linear Unit
ROI	Region Of Interest
SCSM	Spatial Contextual Superpixel Model
SIFT	Scale-Invariant Feature Transform
SOM	Self-Organizing Map
SVM	Support Vector Machine
SVR	Support Vector Regression
TL	Tree Leaf
TS	Tree Stem
VI	Vegetation Index
VocANN	VOCGP obtained using ANN
VocCNN	VOCGP obtained using CNN
VOCGP	Vertical Orientation Connectivity of Grass Pixel
VVI	Visible Vegetation Index

List of Figures

List of Tables

Chapter 1
Introduction

This chapter presents brief background information and datasets for roadside video data analysis. It also shows some relevant applications of video data analysis and presents a detailed outline of this book.

1.1 Background

The roadside is a specific region that is of great significance to many fields such as agriculture, forestry, transportation, power and telecommunications, national security and environmental authorities. Being able to acquire accurate information about roadside conditions can be potentially useful for assisting a wide range of practical applications, such as effective roadside management, vegetation growth condition monitoring, and road hazard assessment. For instance, accurately estimating the site-specific parameters of roadside vegetation such as biomass, height, coverage, density and greenness can be critically important for monitoring the real-time growth conditions of roadside grasses and trees. This supports the design and implementation of effective vegetation management strategies by relevant government authorities, preserving the road assets and improving the safety of the driving public. Tracking the changes in these parameters is an effective way to detect and quantify the possible effects on vegetation such as diseases, dryness, soil nutrients and water stress. From the perspective of safety, roadside vegetation (e.g. grass and tree) with high biomass can present a major fire hazard to the security of drivers and vehicles, particularly in remote and low-population regions without regular and frequent human checks on roadside grass growth conditions.

The current practice of monitoring roadside conditions by relevant authorities is heavily dependent on manual measurements by humans, which suffer from high investments in terms of labor, time, effort and cost. The necessity and importance of developing automatic roadside video data analysis techniques have been widely

B. Verma et al., *Roadside Video Data Analysis: Deep Learning*,
Studies in Computational Intelligence 711, DOI 10.1007/978-981-10-4539-4_1

accepted by both government authorities and academics. Unfortunately, there has been very limited progress in this direction. In fact, the vast majority of current computer scientists working on video data analysis are focusing on developing robust or real-time algorithms that can improve the performance on commonly used benchmark datasets with general types of objects and scenes. Little attention to roadside analysis may be due to the lack of such benchmark datasets, and also due to its restriction to only a relatively small number of specific roadside objects.

In light of the current lack of work specifically focusing on the roadside, this book aims to be one of the first studies that specifically focuses on roadside video data analysis using various modern machine learning algorithms, particularly deep learning techniques, with an emphasis on the use of these techniques for supporting real-world roadside applications. We describe several industry datasets collected from the state roads within the Fitzroy region, Central Queensland, Australia, as well as several commonly used real-world benchmark datasets. We then present an overview of a general systematic framework for roadside video analysis and introduce the most widely used feature types and object classification algorithms in present studies. Various types of non-deep and deep learning algorithms for video analysis are further described in detail and evaluated using industry and benchmark datasets. A case study is specifically introduced regarding the development of automatic and efficient methods to estimating the biomass of roadside grasses and identifying fire-prone road regions. Finally, we discuss several issues and challenges associated with analyzing roadside video data, and highlight possible future directions.

1.2 Collection of Roadside Video Data

A robust and reliable machine learning technique, including deep learning algorithms, is expected to be consistently effective on both local roadside datasets collected for a specific purpose and general scene datasets taken from various real-world scenarios. To enable direct and fair comparisons of the techniques described in this book with other state-of-the-art approaches, this section introduces four industry datasets and five widely used benchmark datasets.

1.2.1 Industry Data

The datasets used for evaluating machine learning techniques in this book come from the roadside video data collected by the Department of Transport and Main Roads (DTMR) using vehicle-mounted cameras in the Fitzroy region, Central Queensland, Australia. Four cameras are mounted on the front, left, right and rear directions of the vehicle, which runs across main state roads in Queensland each year. All video data is in an AVI format with a frame resolution of 1632 × 1248 pixels at approximately 10 meters apart, totaling more than 500 GB. The data from the left-view camera primarily focuses on roadside vegetation regions, and thus can be used to monitor vegetation growth conditions and find potential fire-prone areas, whereby steps can be taken subsequently to eliminate the fire hazard.

Like similar real-world video datasets, the DTMR video data lacks pixel-wise annotation of ground truths of all objects that are present in video frames, which would require intensive labor efforts considering the huge number of all videos and the high resolution of frames. However, ground truths are critically important for designing and evaluating machine learning algorithms, as well as promoting future research efforts in this field. In this book, we document four image datasets based on the video captured using the left or frontal camera.

1.2.1.1 Cropped Roadside Object Dataset

We manually cropped a total of 650 small regions from 230 frames for seven types of objects (100 regions per object except for sky which has 50 regions), including brown grass, green grass, tree leaf, tree stem, soil, road, and sky as shown in Figs. 1.1 and 1.2. All frames were selected from the video captured by the left camera, and thus they primarily focus on the vegetation area rather than the road area. To simulate realistic conditions as closely as possible, the frames were selected to cover a wide range of vegetation types, and different scene content, day times and locations, etc. A set of local regions was then cropped from these frames and each region was ensured to belong to only one object. It should be noted that the resulting rectangular regions have different resolution and shapes. The cropped regions represent various types of appearance of roadside objects, and have very challenging characteristics in terms of variations in the appearance and content within the same object, as well as between objects, making them suitable for evaluating the performance of machine learning algorithms in natural conditions.

Fig. 1.1 Illustration of small regions cropped from DTMR video frames. The *white blocks* in the figures give an indication of the location, size and shape of the cropped regions. Note that only a proportion of representative cropped regions are included in the final cropped roadside object dataset

Fig. 1.2 Samples of cropped regions for seven roadside objects. There are big variations in the appearance of the same object and high similarity between objects (e.g. green grass and tree leaf), which are challenges for accurate object classification. These regions are rescaled to the same shape and size only for illustration

1.2.1.2 Cropped Grass Dataset

The cropped grass dataset was created to evaluate the performance of classifying sparse versus dense grasses, and it comprises 110 color images including 60 for dense grasses and 50 for sparse grasses. The images taken are color images under natural lighting conditions. To cover various types of grasses in real-world situations, different canopy sizes were selected as shown in Fig. 1.3. All images were stored in a JPEG format with a resolution of 900 $\times$ 500 pixels.

Fig. 1.3 Sample images for (*top*) sparse and (*bottom*) dense grasses

1.2.1.3 Natural Roadside Object Dataset

We manually selected 50 images from left-view video data as shown in Fig. 1.4 (independent from those used for the cropped roadside object dataset). These images were selected to be representative of roadside scenes in different real-world cases, covering various types of vegetation and other objects, such as soil, road, and sky. All pixels were manually annotated by a researcher in the computer vision field into six categories of objects, including brown grass, green grass, tree, soil, road,

Fig. 1.4 Sample images in the natural roadside object dataset (*left*) with their corresponding pixel-wise ground truths (*right*) for six types of roadside objects

and sky, and they serve as ground truths for performance evaluations. The uncertain regions were labeled as *unknown*. Unlike the cropped roadside object dataset, two categories of tree leaf and stem were combined into one category of tree, due to relatively small sizes of tree stems in the images.

1.2.1.4 Natural Road Object Dataset

The natural road object dataset was generated to evaluate the performance of object detection in frontal view road conditions. More than 400 images were extracted from DTMR video sequences that were captured using a frontal view camera. The images were taken under different lighting conditions and natural settings, rescaled to a resolution of 960 × 1280 pixels, and stored in a JPEG format. The images primarily focus on road lanes, but also include other objects such as tree, grass, sky and traffic sign. The presence of three objects, including road, sky and traffic sign, was manually annotated to serve as ground truths for road object detection. A sample set of images used is shown in Fig. 1.5.

Fig. 1.5 Sample images in the natural road object dataset. The images were captured using a frontal view camera

1.2.2 *Benchmark Data*

1.2.2.1 The Stanford Background Dataset

The Stanford background dataset [1] comprises of 715 images of outdoor scenes assembled from existing public datasets, including LabelMe, MSRC, PASCAL and Geometric Context. There are eight object classes, including sky, tree, road, grass, water, building, mountain and foreground object. Images are approximately 320 × 240 pixels, and each contains at least one foreground object. All image pixels were manually annotated into one of eight classes or unknown object using Amazon Mechanical Turk. A set of sample images is shown in Fig. 1.6.

1.2.2.2 The MSRC 21-Class Dataset

The MicroSoft Research Cambridge (MSRC) 21-class dataset [2] was designed to evaluate multi-class object segmentation or recognition. As shown in Fig. 1.7, it contains 591 images labeled with 21 classes, including building, grass, tree, cow, sheep, sky, airplane, water, face, car, bicycle, flower, sign, bird, book, chair, road, cat, dog, body, and boat. The images have a resolution of 320 × 213 or 320 × 240 pixels. The dataset provides an approximate pixel-wise ground truth labelling of class categories, and a *void* label to cope with pixels that do not belong to one of the 21 classes.

Fig. 1.6 Sample images and their corresponding pixel-wise ground truths in the Stanford background dataset. The images were captured in a wide range of real-world situations, and form one of the most challenging datasets in the field of scene content understanding

Fig. 1.7 Sample images and their corresponding pixel-wise ground truths in the MSRC 21-class dataset. Note that *black pixels* in the ground truth images indicate a *void* label, which is excluded from the performance evaluation

1.2.2.3 The SIFT Flow Dataset

The SIFT Flow dataset [3] includes 2688 images that were thoroughly labelled by LabelMe users. Most of the images are outdoor scenes and the top 33 object categories with the most labeled pixels are included in the dataset, including sky, building, mountain, tree, road, beach, field, etc. There is also an additional *unlabeled* category for pixels that are not labelled or labelled as other object categories. The images have a resolution of 256×256 pixels, and a set of sample images is demonstrated in Fig. 1.8. It is worth noting that most existing studies on using the SIFT Flow dataset follow the train/test data split evaluation strategy in [3]: 2488 training and 200 test images.

Fig. 1.8 Sample images and their corresponding pixel-wise ground truths in the SIFT flow dataset. Note that there are 33 objects in total, but only a subset of object categories is displayed in the sample images

1.2.2.4 The Croatia Roadside Grass Dataset

The Croatia roadside grass dataset includes 270 images randomly extracted from video recordings captured using a right-view HD Camcorder Canon XF100 camera from a vehicle moving along public roads. The recordings were collected at different day times during the day and include different traffic scenarios including roadside vegetation in various conditions. The images were selected to cover vegetation in different lighting conditions, mowed and uncut grass, bushes, and objects of color similar to vegetation. Focusing primarily on green grasses and roads as shown in Fig. 1.9, the images have a frame resolution of 1920 × 1080 pixels and image pixels have been manually annotated into one of two categories—grass and non-grass.

1.2.2.5 The MNIST Dataset

The Mixed National Institute of Standards and Technology (MNIST) dataset of handwritten digits is widely used in evaluating pattern recognition algorithms. It is a subset of a larger set available from the NIST. There are 10 digits from 0 to 9,

Fig. 1.9 Sample images and their corresponding pixel-wise ground truths in the Croatia roadside grass dataset

Fig. 1.10 Samples of handwritten digits in the MNIST dataset (adopted from [4])

which have been size-normalized and centered in 28 × 28 grey images as shown in Fig. 1.10. The dataset contains 70,000 handwritten patterns, including 60,000 training and 10,000 test samples. The 60,000 pattern training set contains examples from approximately 250 writers. The dataset is publicly available at the site http://yann.lecun.com/exdb/mnist/.

1.3 Applications Using Roadside Video Data

There are many potential applications that can benefit from roadside video data analysis. This section describes several typical applications that play a significant role in real-world applications.

(1) Roadside vegetation status monitoring. Accurately estimating the site-specific parameters of roadside vegetation such as biomass, height, coverage and density plays an important role in many applications such as assisting growth condition monitoring and roadside vegetation management. These parameters can provide reliable and important indications of the current condition, growth stage and future tendency of vegetation. Tracking the changes in these parameters is an effective way to detect and quantify the possible effects on the vegetation such as diseases, dryness, soil nutrients and water stress, and accordingly implement suitable measurements to achieve the expected cultivation goals in agriculture and forestry.

(2) Roadside fire risk assessment. Roadside vegetation such as grasses and trees with high biomass can present a major fire hazard to the safety of drivers and vehicles, particularly in remote and low-population regions without regular and frequent human checks on roadside grass growth conditions. Due to an extremely large number and unpredictable complex conditions of road networks, it is often infeasible and impossible to constantly and frequently visually check on potential fire risks by humans for all roadsides in general practice. The fire hazards from high-biomass vegetation can possibly occur at any location on roadside sites, leading to major threats to the safety of drivers and vehicles. In addition, the fire might also spread to neighboring regions resulting in bigger disasters such as wild fires. Therefore, the development of automatic and efficient methods to estimate the biomass of roadside vegetation is very important for the transport authority in identifying fire-prone road regions and taking necessary actions to burn or cut the vegetation to prevent possible hazards.

(3) Roadside vegetation regrowth management. The branches of some types of vegetation such as trees and bushes can grow up quickly approaching the road shoulder boundary, which might impact normal driving on the road, while other types of vegetation might be growing slowly and far away from the road, imposing little threat. Identifying the regrowth conditions of roadside vegetation such as the closeness to the road, the height, size, and greenness of

vegetation, and implementing approximate measurements to cut those potentially threatening branches are also critical in ensuring safe driving conditions. In current practice, such identification is often completed by humans, and thus developing automatic techniques to monitor the regrowth condition of roadside vegetation can be very useful in freeing human efforts and reducing associated costs. The results can also assist relevant transport authorities decide what types of equipment should be used to manage the regrowth conditions on roads; e.g. small equipment can be used to cut small trees and shrubs, while big equipment for big trees. This is also useful for estimating the costs, as big trees are more expensive to be removed, while small shrubs impose relatively small expenses.

1.4 Outline of the Book

The rest of this book is organized as follows.

In Chap. 2, we present a roadside video data analysis framework, which highlights the main processing steps and components of a practical roadside video analysis system. It provides a blueprint for employing relevant algorithms and techniques for the purpose of roadside video data analysis. The key techniques used for feature extraction, object segmentation and object classification are also introduced briefly. In addition, we also review related work on vegetation and generic object segmentation. Finally, the Matlab codes of several commonly used data processing algorithms are provided.

In Chap. 3, we describe in detail various types of non-deep learning techniques that have been used or can be potentially used for roadside video data analysis. We introduce the main processing steps in each type of techniques and conduct experiments to evaluate their performance on both industry and benchmark datasets.

In Chap. 4, we introduce deep learning techniques and their applications to roadside data analysis. We experimentally compare the automatic feature extraction in Convolutional Neural Network (CNN) with hand engineering features based on the same Multi-Layer Perceptron (MLP) classifier, as well as compare an assemblage architecture of CNNs with a single classifier model of CNN or MLP. We then propose a novel deep learning network for object segmentation, and demonstrate the state-of-the-art performance on three benchmark datasets.

In Chap. 5, we present a case study that develops a systematic framework for roadside fire risk assessment utilizing machine learning techniques. We experimentally compare features automatically extracted by CNN with hand engineering texture and color features in the estimation of the fuel load of roadside grasses using a Vertical Orientation Connectivity of Grass Pixel (VOCGP) algorithm. We describe the field sample data with objective fuel loads collected from a list of roadside sites in central Queensland regions, and then show promising performance of the framework in identifying fire-prone regions across state roads in the regions.

In Chap. 6, we provide some recommendations to work in this direction, and discuss possibly future challenges and opportunities for promoting research and applications on utilizing roadside data analysis techniques.

References

1. S. Gould, R. Fulton, D. Koller, Decomposing a scene into geometric and semantically consistent regions, in *IEEE 12th International Conference on Computer Vision (ICCV)*, 2009, pp. 1–8
2. J. Shotton, J. Winn, C. Rother, A. Criminisi, Textonboost for image understanding: multi-class object recognition and segmentation by jointly modeling texture, layout, and context. Int. J. Comput. Vis. **81**, 2–23 (2009)
3. L. Ce, J. Yuen, A. Torralba, Nonparametric scene parsing: label transfer via dense scene alignment, in *IEEE Conference on Computer Vision and Pattern Recognition (CVPR)*, 2009, pp. 1972–1979
4. Y. Lecun, L. Bottou, Y. Bengio, P. Haffner, Gradient-based learning applied to document recognition. Proc. IEEE **86**, 2278–2324 (1998)

Chapter 2
Roadside Video Data Analysis Framework

This chapter introduces a general framework for roadside video data analysis. The main processing steps in the framework are described separately. It also reviews previous related work on vegetation and generic object segmentation, and lists several commonly used data processing algorithms.

2.1 Overview

Figure 2.1 depicts a general framework for roadside video data analysis, which is composed of five main steps. For a given roadside video, the data is firstly pre-processed to make it suitable for further processing in the framework. For instance, the video can be converted into a sequence of static frames and rescaled to the same resolution. An object segmentation step is then employed to segment each frame into regions of objects, from which a set of representative features is extracted for each object of interest and further used for the classification of the states of the object (e.g. low vs. high fuel load of grasses) using deep learning or other machine learning algorithms. Once all objects of interest are correctly classified, the applications can be implemented accordingly based on the goals of a specific application.

It is important to note that designing automatic systems for vegetation segmentation and classification in natural video data generally face many challenges, such as unstructured, dynamic or even unpredictable configuration of vegetation, significant changes in environmental conditions, and high dependence on data capturing settings such as camera configuration and resolution. The scene may be also overexposed, underexposed or blurred. Therefore, it is critical to take all or a part of these challenges into consideration when designing techniques for different processing steps of the whole system.

B. Verma et al., *Roadside Video Data Analysis: Deep Learning*,
Studies in Computational Intelligence 711, DOI 10.1007/978-981-10-4539-4_2

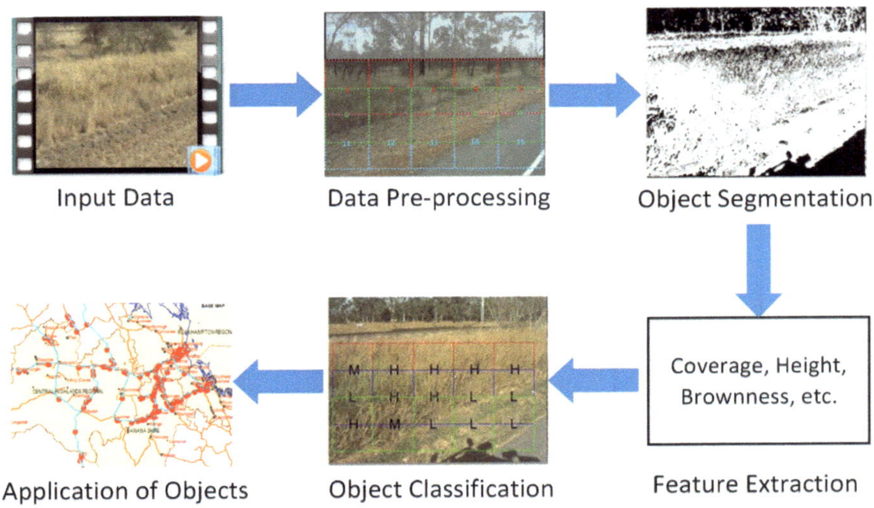

Fig. 2.1 A general framework for roadside video data analysis

2.2 Methodology

2.2.1 Pre-processing of Roadside Video Data

Data pre-processing is an important step in providing suitable types of data for the steps that follow and ensuring processing outcomes as expected. Depending on the purpose of each specific application and the nature of the video data, different types of pre-processing techniques can be employed in this step. Here, we introduce several commonly used pre-processing techniques.

(1) Video to frame conversion. Without neglecting the fact that there are many applications that demand only raw video data and directly perform processing on them to obtain the desired outcomes, it is often the case that a conversion step from video data to a sequence of frames is a pre-requisite for many applications. Extracting frames from the input video data allows performing further detailed analysis and processing on each individual frame separately, which is often vital in obtaining detailed information as required by specific applications. Figure 2.2 exemplifies the processing of extracting a series of frames from a roadside video.
(2) Color space conversion. There are many types of color spaces that have been proposed previously which have their own characteristics in the perception of the video frame content, including RGB, HSV, CIELab, $O_1O_2O_3$, etc. For robust feature extraction, it is a common step to convert a color space into another suitable space (e.g. [1]) that is more robust or invariant to effects from the environment such as shadow, illumination, and dynamic and uncontrollable

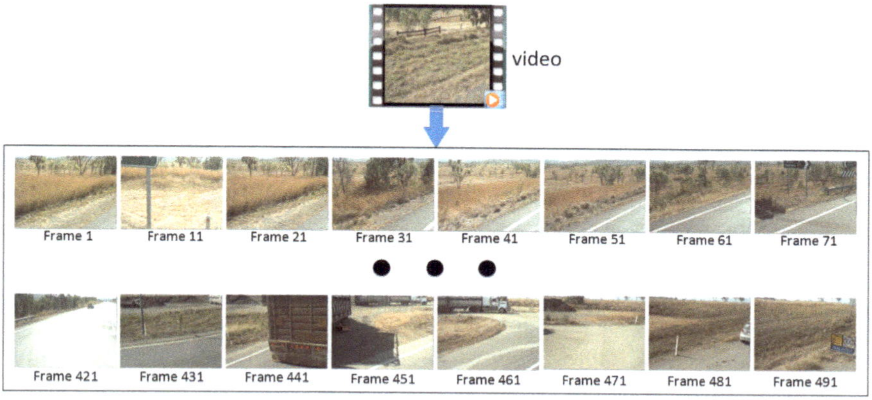

Fig. 2.2 An example of video to frame conversion. There are 500 frames in total in the example video, and every 10th frame is displayed

lighting conditions. For algorithms such as Gray-Level Co-occurrence Matrix (GLCM) and histogram equalization that require grayscale images, the original color images are needed to be converted to grayscale. RGB to gray conversion is one of the most frequently used conversion methods, as most existing color images can be represented by R, G, and B channels. Among many ways of converting RGB to grayscale, such as taking the mean value over R, G, and B, the most commonly used one is based on Eq. 2.1:

$$I = 0.2989 \times \mathrm{R} + 0.5870 \times \mathrm{G} + 0.1140 \times \mathrm{B} \tag{2.1}$$

where, *R*, *G*, and *B* stand for red, green and blue channels, respectively and *I* is the grayscale image. The advantage of this conversion is that it does not treat each color channel equally, which corresponds to the fact that humans' eyes perceive green more strongly than red and red more strongly than blue. Figure 2.3 shows an example of RGB to gray conversion.

Fig. 2.3 An example of converting an RGB image (*left*) to a *grayscale* image (*right*)

(3) Frame resizing. Due to different specifications of video recording or image capturing equipment, video frames may have large variations in the resolution of pixels. For roadside video analysis systems that have requirements of the size of input data and computational time, it becomes a necessary step to resize frames to a desired resolution. A crucial step is to choose a proper rescaled size to accommodate data with different resolutions. Normally, down-scaling the size of frames results in the loss of some information about the content, while up-scaling introduces artificially generated information, which may significant impact the performance of the system. There are also various types of resizing algorithms available for use, such as nearest-neighbor interpolation, bilinear interpolation, and bicubic interpolation, which also have influence on the quality of resized frames. Figure 2.4 shows the results of image resizing on samples from the cropped roadside object dataset.

(4) Histogram equalization. Data captured in real-world environmental conditions may be exposed to different lighting effects, such as shadow, shining, under- and over-exposure. These effects form a major challenge for the robustness of machine learning algorithms, as they may substantially change the appearance of a proportion or all parts of the scene data and lead to confusion between objects. Although many studies [2] have investigated techniques to overcome some of these effects in scene content understanding, it is a common pre-processing step to firstly perform illumination adjustments to ensure even illumination in the scene data before inputting the data into further processing steps. One of most popular approaches for handing uneven illumination is performing histogram equalization, which transforms the intensity image into one with an approximately equally distributed histogram, and thus the lighting effects are reduced.

(5) Noise removal. Image or video data is often prone to a wide range of noise, such as salt and pepper noise, that do not reflect the true intensities of real-world objects. The noise can be introduced at the data acquisition stage, transmission

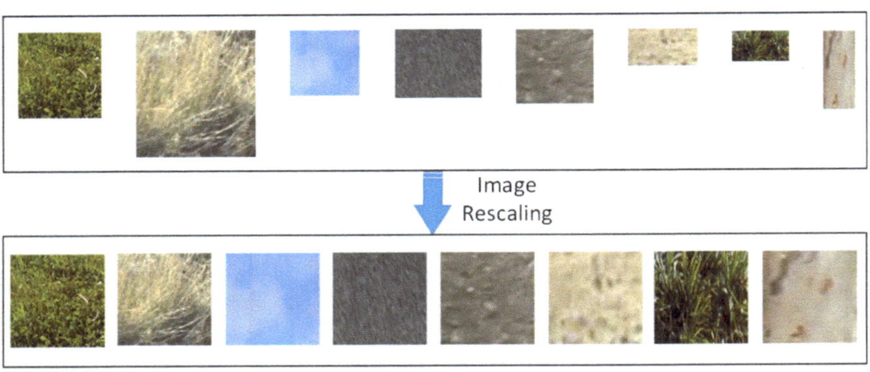

Fig. 2.4 Examples of cropped roadside regions resized into the same resolution by applying an image resizing technique

stage, or post-processing stages, depending on the method used for data creation. There are many image filtering methods for noise removal, such as the averaging filter, medium filter, Sobel filter, and Wiener filter. Figure 2.5 displays an example image with noise removed by applying a median filter.

(6) Sample region selection. Within the whole frame captured to represent a wide range of scene content and objects, in most cases, only a proportion of the frame is of great interest to the end user or a specific application. Thus, it becomes necessary to perform sample region selection to obtain the Region of Interest (ROI) from the scene that corresponds precisely to the roadside region used in the practice. The region selected is often different for different applications and this selection process can often be assisted by manual cropping by human or pre-setting the location, size, and shape of the ROI in an automatic system. Figure 2.6 shows an example of a sample region in field tests and its corresponding sample region in the captured image.

Fig. 2.5 An example of noise removal (*right*) by applying a median filter of 3 × 3 pixels to an original roadside image (*left*)

Fig. 2.6 A sampling region indicated by a *white plastic square* in field tests (*left*) and one possible corresponding sample region indicated by a *red rectangle* in the image (*right*)

2.2.2 Segmentation of Roadside Video Data into Objects

Object segmentation aims to find out the type of each possible object and the location where it is present in the video or frame data. Object segmentation is a pre-requisite step in many computer vision tasks, and it supports detailed analysis or further processing on the objects of interest. Object segmentation itself is a relatively popular research direction with numerous studies in the literature from the perspectives of scene labelling, scene parsing, image segmentation, etc. However, automatic and accurate segmentation of vegetation from natural roadside data is still a challenging task, due to substantial variations in both the unconstrained environment and the appearance of objects. The data may be accompanied with various outdoor environmental effects, such as overexposure, underexposure, shadow and sunlight reflectance. It is still a difficult task to accurately predict the type and appearance of objects that are present in a new scene, even with prior knowledge about the location, season, time, weather condition, etc. Figure 2.7 shows two roadside frames and their corresponding object segmentation results.

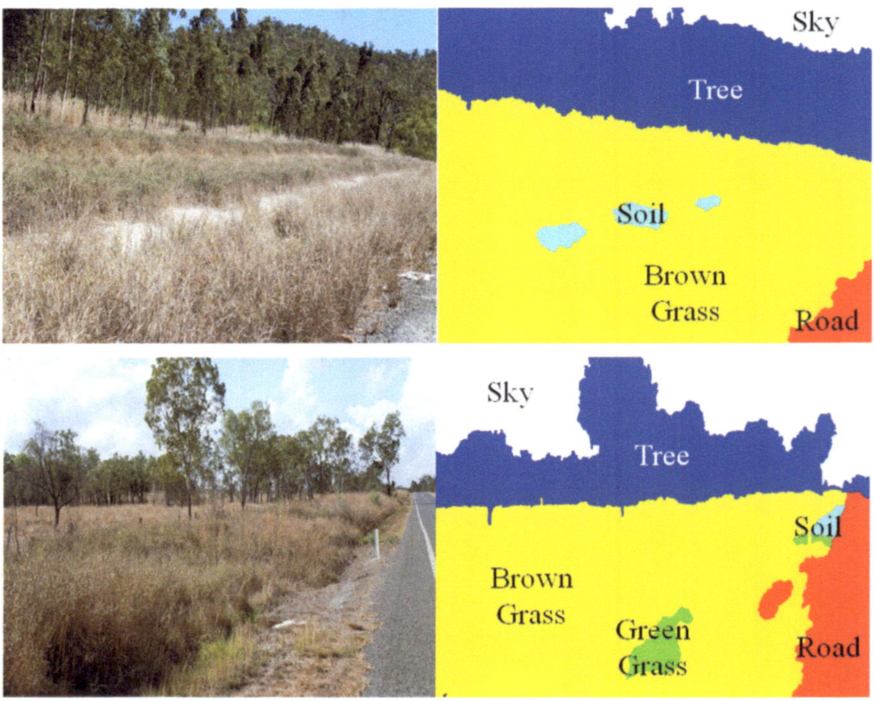

Fig. 2.7 Graphic illustration of object segmentation in roadside frames. The frames are segmented into different object categories, such as tree, sky, road, soil, brown grass, and green grass

2.2.3 Feature Extraction from Objects

To be able to recognize objects, it is often a pre-requisite to extract a set of features that can effectively represent visual characteristics of different objects. Dependent on the aim of a specific application, different types of features need to be extracted. For instance, sky can be generally represented in a blue or while color, while trees are primarily featured by a green or yellow color. It should be noted that the features can be used for both object segmentation and object classification tasks. Challenges for automatic feature extraction are changes in the color and intensity of light in different sunshine conditions and outdoor environments, as well as the lack of specific shapes and texture for most types of vegetation. Most recent studies reporting successful segmentation of vegetation have focused on features extracted from specific species of vegetation but not general vegetation.

According to the optical spectrum of data capturing equipment, features used in existing studies can be roughly divided into two categories: visible features and invisible features.

(1) Visible features reflect the shape, texture, geometry, structure and color characteristics of roadside objects such as sky, road and soil in the visible spectrum. These features are often extracted in the visible spectrum, and thus have high consistency with human eye perception. Color is one of the dominant resources that the human eyes depend on in the perception and discrimination of different types of objects. Some vegetation do not have a specific type of texture or shape, but usually can be represented by a dominant color. The usual color channels presented in vegetation regions include green, red, orange, brown and yellow, and the most popular color spaces include RGB, HIS, HSV, YUV, and CIELab. However, there are also objects that share similar color characteristics and thus cannot be easily distinguished solely using color features, specifically in complicated real-world environmental conditions. For instance, vegetation color is believed to be green in the HSV space under most environmental conditions. However, this may not be the case in scenes containing sky and with varying lighting conditions, such as the presence of shadow, shining, under- and over-exposure effects. For objects that are difficult to be discriminated by color features, other types of features such as texture, location, and geometric properties are able to provide complementary information and become crucially important for robust classification of those objects. Thus, it is advisable to fuse multiple types of features for better results in natural conditions. Examples of texture features used in the field of computer vision include Local Binary Patterns (LBPs), Gabor filter, Scale-Invariant Feature Transform (SIFT), Histograms of Oriented Gradients (HOGs), and GLCM.

(2) Invisible feature approaches extract the reflectance characteristics of vegetation in the invisible spectrum to differentiate them from other objects. It is widely known that vegetation needs to use chlorophyll to convert sunlight radiant energy from the sun into metabolic energy, which exhibits unique absorption characteristics of wavelengths. Based on this, various types of Vegetation

Indices (VIs) have been designed to characterize the differences between the spectral properties of vegetation and those of other objects on the available bands, especially for green and near infrared wavelengths. One great advantage of invisible features is that they often retain a high robustness against large variations in environmental conditions, such as illumination changes and light exposure, and thus they are suitable for achieving consistently stable performance in real-world environments. By contrast, one major drawback of invisible features is that they require specialized data capturing equipment, such as LIght Detection And Ranging (LIDAR), near-infrared cameras, and sensors. This requirement, to some extent, limits direct adoption of invisible features in a wide range of applications. It still remains a question as to how to define VIs capable of being adapted to any kind of natural conditions. Table 2.1 shows a list of different types of features used in existing studies.

2.2.4 *Classification of Roadside Objects*

Classification of objects aims to recognize the type or state of objects in roadside data, such as fuel load of grass, height of tree, content of traffic sign, and width of road. Given the feature sets extracted for each object, the key task is to design a proper machine learning algorithm that is capable of robustly predicting the state of objects of interest. Although humans are able to easily identify the state of some objects without being impacted by environmental effects such as shadows of objects, the use of machines for automatic object classification is still a challenging task. In the literature, there are a wide range of algorithms and they can be generally

Table 2.1 Typical types of features used in existing studies for video data analysis

Category	Sub-category	Feature
Visible	Color space	Lab, RGB, HSV, YUV, etc.
	Color statistics	Histogram, mean, standard deviation, max, min, variance, entropy, etc.
	Texture	LBP, SIFT, HOG, Gabor filter, GLCM, CWT, Pixel Intensity Difference (PID), etc.
	Geometry	Location, size, shape, area, centroid, eccentricity, etc.
	Motion	Optical flow
	VI	Excess Green (ExG), Excess Red (ExR), Visible Vegetation Index (VVI), Color Index of Vegetation Extraction (CIVE), etc.
Invisible	–	Normalized Difference Vegetation Index (NDVI)
	–	Near Infrared Ray (NIR)
	–	Modification of NDVI (MNDVI)
	–	Laser reflectivity

Note Vegetation Index (VI) can be either invisible or visible features classified into supervised learning and unsupervised learning, which are briefly introduced below.

(1) Supervised learning, which normally involves the design of a suitable machine learning algorithm and finds the optimal parameters of the algorithm based on a labeled training dataset. Each sample in the dataset is composed of an input object and a desired output value, and the whole dataset is often divided into training, validation and test subsets. The parameters of the algorithm are firstly trained using the training subset, further evaluated using the validation subset, and finally applied to classify the state of objects on the test subset, producing performance measures such as prediction accuracy. The most widely used supervised learning based techniques include Artificial Neural Network (ANN), Support Vector Machine (SVM), decision tree, random forest, non-linear regression, Conditional Random Field (CRF), and nearest neighbor algorithm.
(2) Unsupervised learning, which performs predictions without the need of the training data, and directly draws inferences from the dataset with or without labelled ground truths. Unlike supervised learning where a training data subset should be prepared and pre-labelled with ground truths, unsupervised learning tries to infer a prediction function that can best describe the pattern information in the data that is labeled or unlabeled, and thus it has the advantage of not requiring labelling work which is crucially important for applications where manual annotation of ground truths is difficult or even impossible in practice. The most widely unsupervised algorithms include K-means clustering, hierarchical clustering, principal component analysis, independent component analysis, non-negative matrix factorization, and the Self-Organizing Map (SOM).

2.2.5 Applications of Classified Roadside Objects

Once roadside objects are classified, there are a lot of potential applications that can be generated that play an important role in specific areas, such as agriculture, transport, road safety, and natural disaster prevention. This section lists several examples of such applications.

(1) Traffic sign detection. Traffic signs are one of the most important signals in regulating and guiding vehicle drivers for safe driving. Automatic traffic sign detection could be useful in assisting drivers making correct driving decisions, especially in severe weather conditions and for signs that are not visible enough. It is also crucial in developing automatic guidance vehicles with smart sensing capacity that can drive automatically in various road conditions, or that can send alerts to drivers to avoid possible accidents.
(2) Fire-prone region identification. Roadside fire risk arising from roadside vegetation, such as brown grasses and trees, is a major hazard to road safety and can potentially be a contributory factor to major disasters, such as bushfires.

The current practice of transport authorities is heavily dependent on human visual checks to find fire-prone roadside regions, and they still lack effective systems to automatically identify fire-prone regions. The implementation of automatic techniques to handle this issue has become increasingly important, and the investigation of robust machine learning algorithms for roadside object classification can bring us a step closer to this goal. Figure 2.8 shows examples of roadside frames with high or low fire risk.

(3) Roadside vegetation management. Effective roadside vegetation management requires dynamically, accurately, and constantly monitoring of the growth conditions of different species of roadside vegetation. Being able to obtain vegetation species along specific road sites in a certain season can help farmers and agricultural professionals make better decisions and more effective plans on the necessary treatments to ensure health conditions of vegetation and eliminate possible obstacles, such as insects and dry weather.

(4) Roadside tree regrowth control. In some roadside regions, trees can grow progressively approaching the boundary of the road, and thus can potentially pose a hazard to road safety. It is therefore necessary to implement automatic methods to identify the conditions of these potentially dangerous trees and take appropriate actions accordingly to eliminate the potential hazard. There are generally four levels of regrowth conditions, including heavy, medium, light and zero, as shown in Fig. 2.9. A 'zero' level indicates big trees which are far

Fig. 2.8 Examples of roadside grasses with high and low fire risk

Fig. 2.9 Examples of four levels of tree regrowth conditions, including heavy, medium, light and zero

away from the road, while a 'heavy' level implies small trees and shrubs which are close to the road. Normally, trees within 10 meters are considered as being close and dangerous. The results can assist service providers in deciding the right equipment for managing the regrowth conditions at the right places. Small equipment is often required to cut small trees and shrubs, while big equipment for large trees. This is also useful for estimating the cost, as removal of large trees is expensive, while small shrubs require relatively less expense. The aim of tree growth control is to effectively handle regrowth problems, minimize road deficiency, and reduce associated liability.

2.3 Related Work

In this section, we review related work on vegetation segmentation and classification. In addition, we also briefly review existing approaches to segmenting objects from generic scenes, which can be potentially used for vegetation segmentation. It is noted that most of existing work related to vegetation segmentation are from the fields such as remote sensing [3] and ecosystems which use different types of sensors, laser scanners, radar and special types of autonomous vehicles. This section limits the focus to reviewing only those approaches that utilize ground data collected using ordinary digital cameras.

2.3.1 Vegetation Segmentation and Classification

According to the type of features used, existing studies on vegetation segmentation and classification can be roughly grouped into three categories: visible feature approach, invisible feature approach, and hybrid feature approach.

2.3.1.1 Visible Feature Approach

Visible feature approaches attempt to distinguish vegetation from other objects, such as soil, tree, sky and road, by exploring their discriminative characteristics in the visible spectrum, such as color, shape, texture, geometry and structure features. A major advantage of using visible features is that they retain high consistency with human visual perception of objects.

Color is one of the dominant information that the human eyes depend on in the perception and discrimination of different objects in real-world environments. Most vegetation are characterized primarily by a green or orange color, and thus color is one of most widely used features in existing studies on vegetation segmentation, which mainly focus on investigating the suitability of various color spaces, such as CIELab [4], YUV [5], HSV [6], and RGB [7]. However, it is still a challenging task to find a suitable color representation of vegetation in complex natural conditions. Designing color spaces that are illumination invariant or able to automatically adapt to the dynamically changing environment is still an active research direction [1].

Except for color, another popular type of visible feature is texture, which mainly reflects the appearance structure of objects and is often represented by performing wavelet filters, such as Gabor filters [8] and Continuous Wavelet Transform (CWT) [6], extracting pixel intensity distributions, such as Pixel Intensity Differences (PIDs) [4, 5] and variations in a neighborhood [9, 10], or generating spatial statistic measures [10], entropy [7], or statistical features over superpixels [11].

Table 2.2 lists typical visible approaches for vegetation segmentation in existing studies. One of the early studies on vegetation segmentation in outdoor images was presented in [12], which employed an SOM for object segmentation and then extracted color, texture, shape, size, centroid and contextual features of segmented regions for 11 object classification using a Multi-Layer Perceptron (MLP). In [7], the entropy was used as a texture feature, together with RGB color components and an SVM classifier for detecting vegetation from roadside images. The intensity differences between pixels were combined with a 3D Gaussian model of YUV channels for grass detection [5], and with L, a, and b color channels for object segmentation [4]. The motion between video frames estimated by optical flow was also employed in a pre-processing step to detect a ROI [6], from which color and texture features were extracted using a two-dimensional CWT, and also to assist vegetation detection by measuring the resistance of vegetation [13]. In [14], LBP and GLCM were combined for discriminating between dense and sparse roadside

Table 2.2 Summary of typical visible approaches for vegetation segmentation

Ref.	Color	Texture	Classifier	Object	Data	Acc. (%)
[12]	RGB, O1, O2, R − G, (R + G)/2 − B	Gabor filter, shape	SOM + MLP	Veg, sky, road, wall etc.	3751 R	61.1
						80
[7]	RGB	Entropy	RBF SVM + MO	Veg versus non-veg	270 I	95.0
[6]	RGB, HSV, YUV, CIELab	2D CWT	SVM + MO	Veg versus non-veg	270 I	96.1
[5]	YUV (3D Gaussian)	PID	Soft segmentation	Grass versus non-grass	62 I	91
[8]	O1, O2	NDVI and MNDVI	Spreading rule	Veg versus non-veg	2000 I	95
		Gabor filter			10 V	
[18]	H, S	Height of grass (ladar)	RBF SVM	Grass versus non-grass	N/A	N/A
[4]	Lab	PID	K-means clustering	Object segmentation	N/A	79
[10]	Gray	Intensity mean and variance, binary edge, neighborhood centroid	Clustering	Grass versus artificial texture	40 R	95
						90
[14]	Gray	LBP, GLCM	SVM, ANN, KNN	Dense versus sparse grass	110 I	92.7
[15]	RGB, HLS, Lab	Co-occurrence matrix	Gaussian PDF + global energy	5, 5 and 7 objects	41 I	89.9
					87 I	90.0
					100 I	86.8
[19]	RGB, Lab	Color moment	Superpixel merging	7 objects	650 I	>90
					50 I	77%

N/A not available, *Veg* vegetation, *non-veg* non vegetation, *I* image, *V* video, *R* region

grasses using majority voting over three classifiers—SVM, ANN, and K-Nearest Neighbour (KNN). In [15], RGB, HLS, and Lab color channels and co-occurrence matrix based texture features were fused for outdoor scene analysis. A set of initial seed pixels was selected based on probabilistic pixel maps which were built using a Gaussian probability density function on a selected subset of color and texture features, and pixels were then grown from the initial seeds by integrating region and boundary information in the minimization of a global energy function.

There are also many studies [16, 17] that investigated detecting or classifying crops from other objects such as soil and weeds using image or video data captured in crop fields. Most of the studies accomplished the recognition task using the green color characteristics of crops, and were often based on simplified environmental conditions rather than natural conditions and thus are not reviewed here.

Most existing visible approaches focus on a binary classification of vegetation versus non-vegetation. Although various types of color and texture features in the visible spectrum can often achieve promising performance, there is no common feature set that is widely accepted as being capable of working well in natural conditions. Most visible approaches become problematic when similar kinds of objects present in the scene like grasses and trees, and green vehicles and green grasses, and when illumination conditions change. An alternative solution is adopting features in the invisible spectrum which are capable of remaining better robustness against environmental variations.

2.3.1.2 Invisible Feature Approach

Invisible feature approaches use the spectral properties of chlorophyll-rich vegetation and their reflectance characteristics in the invisible spectrum to differentiate vegetation from other objects, or to determine their properties (e.g. passable vegetation detection in vehicle navigation [13]). It is well known that vegetation needs to use chlorophyll to convert sunlight radiant energy from the sun into metabolic energy, which exhibits unique absorption characteristics of wavelengths. From this theory, various types of VIs have been designed to highlight the differences between the spectral properties of vegetation and those of others on the available bands, especially for green and near infrared wavelengths. Compared with visible features, one prominent advantage of invisible features is that they have better robustness to environmental variations such as shadow, shining and underexposure effects.

The power of VI for vegetation classification has been demonstrated by the fact that a simple pixel-by-pixel comparison between red and Near Infrared Ray (NIR) reflectance potentially provides a powerful and robust way to detect photosynthetic vegetation [20]. In general, healthy and dense vegetation reflects high NIR and low red reflectance. Conversely, sparse and not so healthy vegetation shows low NIR reflectance but high red reflectance. The NIR has been modified to the Normalized Difference Vegetation Index (NDVI) which was successfully applied to vegetation detection [20]. Under illumination changes, Nguyen et al. [2] found that the hyperplane to classify vegetation and others could be in a logarithmic form instead of a linear one in the standard form of NDVI, and thus they proposed the Modification of NDVI (MNDVI). They also experimentally proved that the MNDVI performs more robust and stable vegetation detection than the NDVI under various illumination effects such as shadow, shining, under- and over-exposure. However, the softening red reflectance impact in MNDVI presents problems in an under-exposure or a dim lighting condition. In contrast, NDVI reveals good performance in those circumstances. Thus, a combination of them was adopted in [13] to achieve more robustness against illumination changes. Wurm et al. [21] measured the remission of laser to classify vegetation and non-vegetation regions in structured environments. Spectral reflectance of vegetation was introduced by Bradley et al. [20] for ground-based terrain classification, and this method was

Table 2.3 Summary of typical invisible approaches for vegetation segmentation

Ref.	Feature	Classifier	Object	Data	Result (%)
[20]	Density, surface normal, scatter matrix eigenvalue, RGB, NIR, and NDVI	Multi-class logistic regression	Veg, obstacle or ground	Two physical environments	95.1
[2]	MNDVI	Threshold	Veg versus non-veg	5000 I, 20 V	91
[13]	MNDVI, NDVI, background subtraction, dense optical flow	Fusion	Passable veg versus others	1000 I	98.4
[22]	Laser reflectivity, measured distance, incidence angle	SVM	Flat veg versus drivable surfaces	36,304 veg 28,883 street	99.9
[23]	Filter banks in Lab and infrared	Joint boost + CRF	Eight classes (road, sky, tree, car, etc.)	2 V	87.3
[9]	Ladar scatter features, intensity mean and std., scatter, surface, histogram	SVM	Veg versus non-veg	500 I	81.5

Veg vegetation, *non-veg* non vegetation, *I* image, *V* video

further improved with better robustness by adopting the laser. It has been shown that, by adding independent light and varying the exposure time, a vegetation detection system is able to perform more robustly against varied illumination conditions [2]. Table 2.3 lists typical invisible approaches for vegetation segmentation in existing studies.

However, one major drawback of invisible feature approaches is that they require specialized data capturing equipment, such as laser scanner, near-infrared camera, and sensor. Thus, a direct adoption of invisible features in various applications is still restricted by this requirement. How to define VIs capable of being adaptive robustly to any kind of natural conditions is still a question in this field?

2.3.1.3 Hybrid Feature Approach

Hybrid feature approaches combine invisible and visible features for more robust and accurate classification results, by utilizing both the capacity of visible features in representing visual appearance of objects and the robustness of invisible features to environmental effects.

Nguyen et al. [8] introduced an active method for a double-check of passable vegetation detection. To calculate statistics features, they used a sliding cube across 3D point clouds, and in each sliding cube, a positive definite covariance matrix was calculated. From the covariance matrix, eigenvalues and eigenvectors were extracted to represent two types of 3D point cloud statistics, including scatter and surface, where scatter represents vegetation such as bushes, tall grasses, and tree canopy, while surface represents solid objects like rocks, ground surface, and tree trunks. However, the 3D features would have difficulty achieving robust vegetation segmentation as they did not consider color information. Therefore, Nguyen et al. [9] proposed a 2D and 3D fusion approach for outdoor automobile guidance with the consideration of color information to detect the location of vegetation areas in the viewed scene. They used six features for training an SVM classifier, including intensity, histograms of color features and 3D scatter features. The intensity features include mean and standard deviation of brightness and color in the HSV space, while the 3D scatter features reflect the spatial structure of vegetation in the local neighborhood of LADAR data. The limitations of this approach lie in the requirement of a long processing time as well as feature values being highly dependent on the environment, the type of sensors, the number of scanned points and the point density. In a similar fashion, Liu et al. [18] combined 2D and 3D features to discriminate between grass and non-grass areas, where the height and color information were obtained from a multi-layer radar and a color camera respectively. The color information is represented by the H and S components in the HSV space. Lu et al. [24] and Nguyen et al. [25] presented approaches combining multiple features e.g. color, texture and 3D distribution information, for vegetation segmentation. In [23], a feature vector from 20-D filter banks was extracted from both visible L, a, b and infrared channels for road object segmentation. A hierarchical bag-of-textons method was then introduced to capture the spatial neighborhood information by extracting multiscale texton features from larger neighboring regions. The textons are essentially centres of each pattern generated using a clustering algorithm. Around 87% global accuracy was achieved for classifying eight objects on the authors' own road scene video dataset. In [8], the opponent color space and Gabor features were combined to measure the similarity between a pixel and its neighbors for vegetation pixel spreading. The NDVI and MNDVI were fused to select chlorophyll-rich vegetation pixels as the initial seed pixels for spreading.

Hybrid feature approaches often produce better results than visible or invisible feature approaches, but they also inherit the drawbacks of both of them such as requiring specialized data capturing equipment. How to select a proper set of visible and invisible features for combination is still another question that should be further explored.

2.3.2 Generic Object Segmentation and Classification

Generic object segmentation and classification approaches aim to find the region where a specific object is present. They often share similar concepts with scene labelling, object categorization, semantic segmentation, etc. in computer vision tasks, and can be potentially applied into roadside vegetation segmentation.

The design of an object classification system has to address several processing tasks, including choosing suitable elementary regions (e.g. pixel, patch and superpixel—groups of neighboring pixels sharing similar appearance and perceptually meaningful atomic regions, as examples shown in Fig. 2.10), selecting discriminative visual features to characterize them (e.g. color, texture and geometry), building robust prediction models for obtaining class label confidence, extracting effective contextual features, and integrating prediction models with contextual information. According to the techniques or features used, existing approaches can be divided into different categories, such as parametric versus non-parametric, supervised versus unsupervised, and pixel based versus region based.

Early approaches to object segmentation obtain class labels for image pixels using a set of low-level visual features extracted at a pixel-level [15, 27] or patch-level [23]. However, because pixel-level features treat each pixel individually, they are unable to capture statistical characteristics of objects in local regions. While patch-level features are able to capture regional statistic features, they are prone to noise from background objects due to the difficulty in accurate segmentation of object boundaries. Recent studies [28–31] have focused more on the use of superpixel-level features as the basic unit for object segmentation, which showed

Fig. 2.10 Visual displays of segmented superpixels in roadside images using a graph-based segmentation algorithm [26]. Different colors indicate different superpixel regions

promising results of extracting discriminative features. Superpixel-level features have several advantages over traditional patch-level features, including being coherent support regions for a single labelling on a naturally adaptive domain rather than on a fixed window, supporting more consistent statistic feature extraction capturing contextual information by pooling over feature responses from multiple pixels, and requiring less computational time. The most commonly adopted superpixel-level features include color (e.g. RGB [32, 33] and CIELab [27, 28, 32, 34, 35]), texture (e.g. SIFT [33, 36], texton [27, 33], Gaussian filter [32], Gist [34] and Pyramid Histogram of Oriented Gradients—PHOGs [34]), appearance (e.g. color thumbnail [33]), location [37], and shape.

Despite the benefits, as visual feature based prediction treats each superpixel independently and does not consider semantic context about the scene, it often faces challenges for object segmentation in complex scenes. For segmentation algorithms, the vast majority of existing approaches focus on graphical models, such as CRF [31] and Markov Random Field (MRF) [33]. Work [31] investigated the use of superpixel neighborhoods for object identification by merging the histogram of a superpixel with neighbors in conjunction with CRF. But the approach still has a high dependence on initial seed superpixel selection. Tighe and Lazebnik [33] obtained superpixel labels using Naive Bayes and utilized a minimization of MRF energy over superpixel labels to enforce contextual constraints of objects. Recently, Balali and Golparvar-Fard [11] extracted a set of superpixel features for the recognition of roadway assets using MRF. Current superpixel based approaches pre-dominantly rely on graphical models (e.g. CRF and MRF), which enforce the spatial consistency of category labels between neighboring superpixels (or pixels) by jointly minimizing the total energy of two items—unary potentials which indicate the likelihood of each superpixel (or pixel) belonging to one of the semantic categories and pairwise potentials which account for the spatial consistency of category labels between neighboring superpixels (or pixels). However, flat graphical models have limited capacity of capturing higher order context.

To overcome this limitation, a wide variety of approaches have been proposed to incorporate contextual information to improve object segmentation accuracy, which is often conducted at two stages: feature extraction and label inference. Feature extraction incorporates the context by designing a rich set of semantic descriptions that represent intrinsic correlations between objects in different types of scenes. The commonly used contextual features include absolute location [27], which captures the dependence of class labels on the absolute location of pixels in the scene, relative location [32], which represents the relative location offsets between objects in a virtually enlarged image, directional spatial relationships [38, 39] which encode spatial arrangements of objects such as beside, below, above and enclosed, and object co-occurrence statistics [36], which reflect the likelihood of two objects co-existing in the same scene. The main drawback of the relative location, directional spatial relationships and object co-occurrence statistics is that they completely discard absolute spatial coordinates of objects in the scene and therefore they cannot capture spatial contextual information such as a high likelihood of sky appearing at the top part of a scene. By contrast, the absolute location excessively retains all pixel coordinates of objects and thus it often demands a large training dataset to collect reliable prior statistics for each object and each image pixel.

Another popular approach of incorporating context is to use graphical models at the label inference stage, such as CRF [29, 40, 41], MRF [42] and energy function [15]. However, these graphical models have two shortcomings: (1) label purity of superpixels cannot be guaranteed due to the difficulty of prefect image segmentation; and (2) only contextual information in local neighborhoods is considered. To handle these shortcomings, one approach is to adopt hierarchical models, such as hierarchical CRF [43], stacked hierarchical learning [34] and pylon model [43], which generate a pyramid of image superpixels and perform classification optimization over multi-levels of images to alleviate the effect of inaccurate region boundaries and utilize higher order contextual information. Another approach is to extract feature descriptions from multiple regions, such as aggregated histograms [31] and weighted appearance features [32]. Although these approaches account for larger context, they still cannot fully capture long-range dependencies of objects in the entire scene and are unable to adapt to scene content.

Another drawback of graphical models is that their parameters are solely learnt from the training data, and thus their performance heavily depends on the availability of adequate training data and they have a generalization issue for new test data. For real-world applications, it is often very difficult or even impossible to collect a large number of training data to ensure adequate training, on top of the fact that it is also extremely time consuming and labor intensive. One solution to this problem, particularly for large datasets, is adopting non-parametric approaches [44], which retrieve the most similar training images to a query image and then perform class label transfer from K-nearest neighbours in the retrieval set to the query image. However, non-parametric approaches still depend on the reliability and accuracy of the retrieval strategies.

Recently, deep learning techniques have shown great advantages in extracting discriminative and compact features from raw image pixels rather than using hand-engineered features. The widely used CNNs utilize convolutional and pooling layers to progressively extract more abstract patterns and demonstrate state-of-the-art performance in many vision tasks [45] including object segmentation. The extracted CNN features can be combined with various classifiers (e.g. MRF, CRF and SVM) to predict class labels. A representative work is by Farabet et al. [29], which applied hierarchical CNN features into CRFs for class label inference in natural scenes. However, the CRF inference is completely independent from CNN training, and thus Zheng et al. [46] formulated the CRF inference as recurrent neural networks and integrated them in a unified framework. In [47], the recurrent CNN feeds back the outputs of CNNs to the input of another instance of the same network, but it works only on sequential data. Recent extensions to CNN models include AlexNet, VGG-19 net, GoogLeNet, and ResNet [48]. However, these models often require adequate image resolutions and may not be directly applicable for roadside vegetation segmentation on datasets such as the cropped roadside object dataset which has lower resolutions and substantial variations in the shape and size.

2.4 Matlab Code for Data Processing

This section introduces several commonly used algorithms for video data pre-processing, feature extraction, object segmentation and classification. The Matlab codes are provided to illustrate the processing steps.

1) Extract all frames from video data.

```
% this code takes as input video data, extracts all frames from the video, and
% saves all framesin a new folder.

% read video data into a variable 'mov'. The path and name of the video are
% represented by the string 'videoFilePathName'.
mov = VideoReader(videoFilePathName);

% set the output folder 'outFrameFold' for storing extracted frames.
frameFolder = outFrameFold;

% create the folder if it does not exist.
if ~exist(frameFolder, 'dir')
   mkdir(frameFolder);
end

% get the total number of frames in the video.
numFrame = mov.NumberOfFrames;

% extract and save each of all frames in a loop.
For iFrame = 1 : numFrame
    % read the ith frame.
    I = read(mov, iFrame);

    % set an index for each frame.
    frameIndex = sprintf('%4.4d', iFrame);

     % write frames with names like 'Frame0001.jpg' into the output folder.
    imwrite(I, [frameFolder 'Frame' frameNewName '.jpg'], 'jpg');

    % show the progress of data processing.
```

```
    progIndication = sprintf('frame %4d of %d.', iFrame, numFrame);
    disp(progIndication);
end
```

2) Convert RGB frames to grayscale frames.

```
% this code converts extracted video frames from a RGB color format to a
% gray scale.

% read frame data into a variable 'I'. The path and name of the frame are
% represented by the string 'imageFilePathName'.
I = imread(imageFilePathName);

% perform frame conversion.
I = rgb2gray(I);

% show converted frames in figures.
imshow(I);
```

3) Frame resizing.

```
% this code takes as input a frame, and resizes the frame into a desired
% width and height.

% perform frame resizing.
I = imresize(I,[numrows numcols]);

% show resized frames in figures.
imshow(I);
```

4) Apply a median filter to a frame.

```
% this code takes as input a frame, and applies a median filter to remove
% noise in the frame.

% convert RGB to grayscale frames.
I = rgb2gray(I);

% perform image filtering using a median filter.
K = medfilt2(I);

% show filtered frames in figures.
imshowpair(I,K,'montage');
```

5) Pixel-level R, G, B feature extraction.

```
% this code takes as input a frame, and extract R, G, and B values at each
% pixel.

% get the dimension of the input frame.
```

```
[numRows, numCols] = size(I);

% scan pixels across all rows and columns of the frame.
for iRow = 1 : numRows
   for iCol = 1 : numCols
         % get R, G and B values at each pixel of the frame.
         pixelRValue = I(iRow, iCol, 1);
         pixelGValue = I(iRow, iCol, 2);
         pixelBValue = I(iRow, iCol, 3);
   end
end
```

6) Patch-level Gaussian feature extraction.

```
% this code takes as input a frame, extracts Gaussian features from a local
% patch centred at each pixel, and stores the resulting features in a 3D matrix.

% set parameters of Gaussian filters.
fixV = 0.7;
size = [7 7];
sigma = 1;
% create Gaussian filters.
fgaus = fspecial('gaussian',size, sigma*fixV);

% get the dimension of the input frame.
[numRows, numCols] = size(I);

% get R, G, and B matrixes separately from the frame.
R = I(:, :, 1);
G = I(:, :, 2);
B = I(:, :, 3);

% apply Gaussian filters to R, G, and B matrixes separately. The resulting
% Gaussian features are stored in a 3D matrix 'filterI'.
filterI (:, :, 1)=conv2(R,fgaus,'same');
filterI (:, :, 2)=conv2(G,fgaus,'same');
filterI (:, :, 3)=conv2(B,fgaus,'same');
```

7) Patch-level statistical feature extraction.

```
% this code takes as input a frame, extracts statistical features from a local
% patch centred at each pixel, and stores the resulting features in variables.

% get R, G, and B matrixes separately from the input frame.
R = I(:, :, 1);
G = I(:, :, 2);
B = I(:, :, 3);
```

```
% set the half size of patches.
nHalfBlock = 4;

% set parameters for handling border pixels.
adHeight = nHeight - nHalfBlock;
adWidth = nWidth - nHalfBlock;
adBegin = nHalfBlock +1;

% scan across all rows and columns of the frame.
for iRow = adBegin:adHeight
      rowBeg = iRow - nHalfBlock;
      rowEnd = iRow + nHalfBlock;
      for iCol = adBegin:adWidth
            colBeg = iCol - nHalfBlock;
            colEnd = iCol + nHalfBlock;

            % get a patch region from which statistical features are extracted.
            patchR = R(rowBeg: rowEnd, colBeg: colEnd);

            % calculate mean, standard deviation, and skewness of R values
            % in the patch region.
            meanPatchR = mean(patchR(:));
            stdPatchR = std(patchR(:));
            skewPatchR = skewness(patchR(:));
      end
end
```

8) Object segmentation and classification.

```
% this code demonstrates the segmentation/classification of objects in
% frames, which accepts a frame as the input and assigns each of all pixels into
% an object category.

% the dimension of the input frame.
[numRows, numCols] = size(I);

% create a variable for storing object categories of all pixels.
objCategory = zeros(numRows,numCols);

% scan across all rows and columns in the frame.
for iRow = 1 : numRows
   for iCol = 1 : numCols
         % get features at each pixel, i.e. RGB values in this example.
         pixelRGBValue = I(iRow, iCol, :);

         % apply a function named 'objSegAlgorithm' to obtain the object
         % category for each pixel based on its features.
```

```
        objCategory(iRow,iCol) = objSegAlgorithm(pixelRGBValue);
    end
end
```

9) Train and test an ANNclassifier.

```
% this code explains the creation and application of a three-layer feedforward
% ANN to classifying new test data.

% create an ANN with parameters of hidden neuron, activation function, and
% learning algorithm.
net = newff(double(TrainX'),double(TrainY'),[15],{'tansig','tansig'},'trainrp');

% parameters controlling the train-test processes of the ANN.
% generate command-line output.
net.trainParam.showCommandLine = true;
% randomly divide training, validation, and test data subsets.
net.divideFcn = 'dividerand';
% ratios of training, validation, and test data subsets.
net.divideParam.trainRatio = 1;
net.divideParam.valRatio = 0;
net.divideParam.testRatio = 0;
% performance measurement.
net.performFcn='mse';
% performance goal.
net.trainParam.goal=0.0001;
% maximum number of training epochs.
net.trainParam.epochs=500;
% minimum performance gradient.
net.trainParam.min_grad = 1e-8;
% maximum validation failures.
net.trainParam.max_fail = 10;

% train the ANN with the training data and store it in the variable 'net'.
[net,tr] = train(net,double(TrainX'),double(TrainY'));

% prediction accuracy on the training data.
outTrainTag = sim(net,TrainX');
% prediction accuracy on the test data.
outTestTag = sim(net,TestX');
```

References

1. W. Maddern, A. Stewart, C. McManus, B. Upcroft, W. Churchill et al., Illumination invariant imaging: applications in robust vision-based localisation, mapping and classification for autonomous vehicles, in *Proceedings of the Visual Place Recognition in Changing Environments Workshop, IEEE International Conference on Robotics and Automation (ICRA)*, 2014
2. D.V. Nguyen, L. Kuhnert, K.D. Kuhnert, Structure overview of vegetation detection. A novel approach for efficient vegetation detection using an active lighting system. Robot. Auton. Syst. **60**, 498–508 (2012)
3. M.P. Ponti, Segmentation of low-cost remote sensing images combining vegetation indices and mean shift. IEEE Geosci. Remote Sens. Lett. **10**, 67–70 (2013)
4. M.R. Blas, M. Agrawal, A. Sundaresan, K. Konolige, Fast color/texture segmentation for outdoor robots, in *IEEE/RSJ International Conference on Intelligent Robots and Systems (IROS),* 2008, pp. 4078–4085
5. B. Zafarifar, P.H.N. de With, Grass field detection for TV picture quality enhancement, in *International Conference on Consumer Electronics (ICCE), Digest of Technical Papers*, 2008, pp. 1–2
6. I. Harbas, M. Subasic, Motion estimation aided detection of roadside vegetation, in *7th International Congress on Image and Signal Processing (CISP)*, 2014, pp. 420–425
7. I. Harbas, M. Subasic, Detection of roadside vegetation using features from the visible spectrum, in *37th International Convention on Information and Communication Technology, Electronics and Microelectronics (MIPRO)*, 2014, pp. 1204–1209
8. D.V. Nguyen, L. Kuhnert, K.D. Kuhnert, Spreading algorithm for efficient vegetation detection in cluttered outdoor environments. Robot. Auton. Syst. **60**, 1498–1507 (2012)
9. D.V. Nguyen, L. Kuhnert, T. Jiang, S. Thamke, K.D. Kuhnert, Vegetation detection for outdoor automobile guidance, in *IEEE International Conference on Industrial Technology (ICIT)*, 2011, pp. 358–364
10. A. Schepelmann, R.E. Hudson, F.L. Merat, R.D. Quinn, Visual segmentation of lawn grass for a mobile robotic lawnmower, in *IEEE/RSJ International Conference on Intelligent Robots and Systems (IROS)*, 2010, pp. 734–739
11. V. Balali, M. Golparvar-Fard, Segmentation and recognition of roadway assets from car-mounted camera video streams using a scalable non-parametric image parsing method. Autom. Constr. **49**(Part A), 27–39 (2015)
12. N.W. Campbell, B.T. Thomas, T. Troscianko, Automatic segmentation and classification of outdoor images using neural networks. Int. J. Neural Syst. **8**, 137–144 (1997)
13. D.V. Nguyen, L. Kuhnert, S. Thamke, J. Schlemper, K.D. Kuhnert, A novel approach for a double-check of passable vegetation detection in autonomous ground vehicles, in *15th International IEEE Conference on Intelligent Transportation Systems (ITSC)*, 2012, pp. 230–236
14. S. Chowdhury, B. Verma, D. Stockwell, A novel texture feature based multiple classifier technique for roadside vegetation classification. Exp. Syst. Appl. **42**, 5047–5055 (2015)
15. A. Bosch, X. Muñoz, J. Freixenet, Segmentation and description of natural outdoor scenes. Image Vis. Comput. **25**, 727–740 (2007)
16. W. Guo, U.K. Rage, S. Ninomiya, Illumination invariant segmentation of vegetation for time series wheat images based on decision tree model. Comput. Electron. Agric. **96**, 58–66 (2013)
17. F. Ahmed, H.A. Al-Mamun, A.S.M.H. Bari, E. Hossain, P. Kwan, Classification of crops and weeds from digital images: a support vector machine approach. Crop Prot. **40**, 98–104 (2012)
18. D.-X. Liu, T. Wu, B. Dai, Fusing ladar and color image for detection grass off-road scenario, in *IEEE International Conference on Vehicular Electronics and Safety (ICVES)*, 2007, pp. 1–4
19. L. Zhang, B. Verma, D. Stockwell, Spatial contextual superpixel model for natural roadside vegetation classification. Pattern Recogn. **60**, 444–457 (2016)

20. D.M. Bradley, R. Unnikrishnan, J. Bagnell, Vegetation detection for driving in complex environments, in *IEEE International Conference on Robotics and Automation*, 2007, pp. 503–508
21. K.M. Wurm, R. Kummerle, C. Stachniss, W. Burgard, Improving robot navigation in structured outdoor environments by identifying vegetation from laser data, in *IEEE/RSJ International Conference on Intelligent Robots and Systems (IROS)*, 2009, pp. 1217–1222
22. K.M. Wurm, H. Kretzschmar, R. Kümmerle, C. Stachniss, W. Burgard, Identifying vegetation from laser data in structured outdoor environments. Robot. Auton. Syst. **62**, 675–684 (2014)
23. Y. Kang, K. Yamaguchi, T. Naito, Y. Ninomiya, Multiband image segmentation and object recognition for understanding road scenes. IEEE Trans. Intell. Transp. Syst. **12**, 1423–1433 (2011)
24. L. Lu, C. Ordonez, E.G. Collins Jr., E.M. DuPont, Terrain surface classification for autonomous ground vehicles using a 2D laser stripe-based structured light sensor, in *IEEE/RSJ International Conference on Intelligent Robots and Systems (IROS)*, 2009, pp. 2174–2181
25. D.-V. Nguyen, L. Kuhnert, T. Jiang, S. Thamke, K.-D. Kuhnert, Vegetation detection for outdoor automobile guidance, in *IEEE International Conference on Industrial Technology (ICIT)*, 2011, pp. 358–364
26. P. Felzenszwalb, D. Huttenlocher, Efficient graph-based image segmentation. Int. J. Comput. Vis. **59**, 167–181 (2004)
27. J. Shotton, J. Winn, C. Rother, A. Criminisi, Textonboost for image understanding: multi-class object recognition and segmentation by jointly modeling texture, layout, and context. Int. J. Comput. Vis. **81**, 2–23 (2009)
28. S. Gould, R. Fulton, D. Koller, Decomposing a scene into geometric and semantically consistent regions, in *IEEE 12th International Conference on Computer Vision (ICCV)*, 2009, pp. 1–8
29. C. Farabet, C. Couprie, L. Najman, Y. LeCun, Learning hierarchical features for scene labeling. IEEE Trans. Pattern Anal. Mach. Intell. **35**, 1915–1929 (2013)
30. A. Sharma, O. Tuzel, D.W. Jacobs, Deep hierarchical parsing for semantic segmentation, in *IEEE Conference on Computer Vision and Pattern Recognition (CVPR)*, 2015, pp. 530–538
31. B. Fulkerson, A. Vedaldi, S. Soatto, Class segmentation and object localization with superpixel neighborhoods, in *IEEE 12th International Conference on Computer Vision (ICCV)*, 2009, pp. 670–677
32. S. Gould, J. Rodgers, D. Cohen, G. Elidan, D. Koller, Multi-class segmentation with relative location prior. Int. J. Comput. Vis. **80**, 300–316 (2008)
33. J. Tighe, S. Lazebnik, Superparsing: scalable nonparametric image parsing with superpixels, in *European Conference on Computer Vision (ECCV)*, 2010, pp. 352–365
34. D. Munoz, J.A. Bagnell, M. Hebert, Stacked hierarchical labeling, in *European Conference on Computer Vision (ECCV)*, 2010, pp. 57–70
35. R. Socher, C.C. Lin, C. Manning, A.Y. Ng, Parsing natural scenes and natural language with recursive neural networks, in *Proceedings of the 28th International Conference on Machine Learning (ICML)*, 2011, pp. 129–136
36. B. Micusik, J. Kosecka, Semantic segmentation of street scenes by superpixel co-occurrence and 3D geometry, in *IEEE 12th International Conference on Computer Vision Workshops (ICCV Workshops)*, 2009, pp. 625–632
37. L. Zhang, B. Verma, D. Stockwell, S. Chowdhury, Spatially constrained location prior for scene parsing, in *International Joint Conference on Neural Networks (IJCNN)*, 2016, pp. 1480–1486
38. Y. Jimei, B. Price, S. Cohen, Y. Ming-Hsuan, Context driven scene parsing with attention to rare classes, in *IEEE Conference on Computer Vision and Pattern Recognition (CVPR)*, 2014, pp. 3294–3301
39. A. Singhal, L. Jiebo, Z. Weiyu, Probabilistic spatial context models for scene content understanding, in *IEEE Conference on Computer Vision and Pattern Recognition, (CVPR)*, 2003, pp. 235–241

40. D. Batra, R. Sukthankar, C. Tsuhan, Learning class-specific affinities for image labelling, in *IEEE Conference on Computer Vision and Pattern Recognition, (CVPR)*, 2008, pp. 1–8
41. Z. Lei, J. Qiang, Image segmentation with a unified graphical model. IEEE Trans. Pattern Anal. Mach. Intell. **32**, 1406–1425 (2010)
42. R. Xiaofeng, B. Liefeng, D. Fox, RGB-(D) scene labeling: features and algorithms, in *IEEE Conference on Computer Vision and Pattern Recognition (CVPR)*, 2012, pp. 2759–2766
43. V. Lempitsky, A. Vedaldi, A. Zisserman, Pylon model for semantic segmentation, in *Advances in Neural Information Processing Systems*, 2011, pp. 1485–1493
44. F. Tung, J.J. Little, Scene parsing by nonparametric label transfer of content-adaptive windows. Comput. Vis. Image Underst. **143**, 191–200 (2016)
45. L. Zheng, Y. Zhao, S. Wang, J. Wang, Q. Tian, Good practice in CNN feature transfer. *arXiv preprint* arXiv:1604.00133 (2016)
46. S. Zheng, S. Jayasumana, B. Romera-Paredes, V. Vineet, Z. Su et al., Conditional random fields as recurrent neural networks. *arXiv preprint* arXiv:1502.03240 (2015)
47. P.H. Pinheiro, R. Collobert, Recurrent convolutional neural networks for scene parsing. *arXiv preprint* arXiv:1306.2795 (2013)
48. K. He, X. Zhang, S. Ren, J. Sun, Deep residual learning for image recognition. *arXiv preprint* arXiv:1512.03385 (2015)

Chapter 3
Non-deep Learning Techniques for Roadside Video Data Analysis

In this chapter, we describe traditional non-deep learning approaches that are used for roadside video data analysis. Each type of these learning approaches is described separately in each section, which primarily focuses on related prior work, technical details of each approach, experimental design, and performance analysis. We also give a short summary of each learning approach at the end of each section.

3.1 Neural Network Learning

3.1.1 Introduction

Neural networks are one of the most widely used machine learning models for various practical applications, particularly computer vision and image classification tasks. The design of neural networks mimics the way a biological brain solves problems and can track back to as early as 1940s [1]. A typical structure of neural networks is often composed of multiple layers with connections between neurons in neighboring layers, including an input layer, an output layer, and one or multiple hidden layers in the middle part. The prediction capacity of neural networks is completed by learning an optimal set of weights for these connections using algorithms such as the backpropagation algorithm based on the training dataset, and the learnt weights inherently convey representative patterns for specific types of input data. For any given new test data, the learnt neural network acts as a black box to produce prediction outputs of the test data in a form of either continuous values such as class probabilities or discrete object categories. For detailed theories on neural networks and their relationships with deep learning, readers can refer to related publications, e.g. [2].

B. Verma et al., *Roadside Video Data Analysis: Deep Learning*,
Studies in Computational Intelligence 711, DOI 10.1007/978-981-10-4539-4_3

In this section, we describe a neural network learning approach [3] for roadside data analysis and demonstrate its performance for object classification on the cropped roadside object dataset and a small set of natural roadside images. We also compare the performance of the neural network learning approach with popular SVM and nearest neighbor classifiers.

3.1.2 Neural Network Learning Approach

Figure 3.1 depicts the neural network learning approach that fuses color and texture information for roadside object classification [3]. For an input image in the RGB space, it is firstly converted to the opponent $O_1O_2O_3$ color space to achieve more robustness against environmental effects. To represent color information, three $O_1O_2O_3$ color channels are extracted at each pixel, while the first three color moments of each color channel are also extracted from a neighboring patch centered at the pixel to represent texture features. A feature-level fusion of the extracted color and texture features is then performed and fed into a multi-class ANN classifier for classifying six roadside objects, including brown grass, green grass, tree leaf, tree stem, road and soil.

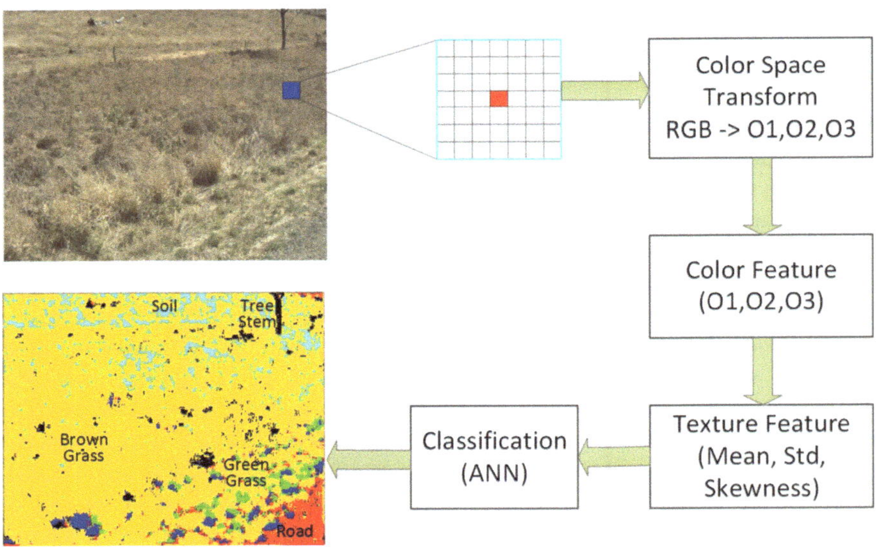

Fig. 3.1 Graphic illustration of the neural network learning approach. For each pixel in an image, O_1, O_2, O_3 color intensity features are extracted. Color moment texture features are also extracted in a local patch centered at the pixel. The color and texture features are combined using a feature-level fusion and fed into an ANN for classifying six object categories

3.1.2.1 Feature Extraction

(1) Color features. Color is one of the most prominent features for object segmentation. For roadside data, most types of tree leaves and green grasses are presented in a green color, while soil is often in a yellow color. Existing studies have used many color spaces, such as HIS, HSV, RGB, and YUV, but there is yet a color space that has been proved as having superior performance than others under different environments. One major problem in using color features is that they are often prone to illumination variations, and thus choosing a suitable color space that has high robustness against environmental effects is a crucial pre-processing step. In this section, we use the opponent color space, which has been shown as having a high level of robustness to illumination changes [4]. The opponent color space has also been successfully used for vegetation detection in outdoor environments [5]. From a RGB color space, the opponent color channels can be obtained using:

$$\begin{bmatrix} O_1 \\ O_2 \\ O_3 \end{bmatrix} = \begin{bmatrix} \frac{R-G}{\sqrt{2}} \\ \frac{R+G-2B}{\sqrt{6}} \\ \frac{R+G+B}{\sqrt{3}} \end{bmatrix} \tag{3.1}$$

In the opponent space, O_1 and O_2 represent the color information, while O_3 represents the intensity information. One advantage of the subtraction in O_1 and O_2 is that it cancels out the light effects equal for all channels and thus they are invariant to light intensity [6]. However, as O_3 reflects only the intensity of objects, it does not have such invariance property. Thus, the color feature vector of a pixel at the coordinate (x, y) in an image is composed of:

$$V^c_{x,y} = [O_1, O_2, O_3] \tag{3.2}$$

(2) Texture features. Roadside objects are not only characterized by color information, but also by texture features. Texture features are better at representing the visual appearance and structure of objects. For objects such as green grasses and tree leaves that have a high level of similarity in the color, it is crucial to incorporate texture features as well. There are a large number of texture descriptors, such as Gabor filters, SIFT features, and filter based textons. However, these descriptors often require the calculation of statistic features (e.g. histograms) in an adequately large region to reliably represent each object category. In natural conditions, this requirement cannot always be met, largely due to a low resolution of captured images or specific nature of the techniques used. Thus, the neural network learning approach employs color moments to represent texture information.

The color moments [7] are the color distribution characteristics in an image or a region. The spatial structure of color conveys important information in representing the appearance of objects. One advantage of the color moments is that they encode both shape and color information of objects, and are invariant to scaling, rotation, and secularities. They have demonstrated excellent performance of object segmentation under varied viewpoints and unstable illumination conditions. Because the spatial structure of color is mostly distributed in the low-order moments, we employ the first three color moments in the opponent color space. For a pixel at (x, y), let P be a patch surrounding the pixel with a size of $N \times N$ pixels, the mean, standard deviation, and skewness of the ith color channel O_i are calculated using:

$$M_{x,y}^{i} = \sum_{j \in P} \frac{1}{N^2} O_i^j \tag{3.3}$$

$$SD_{x,y}^{i} = \sqrt{\frac{1}{N^2} \sum_{j \in P} (O_i^j - M_{x,y}^{i})} \tag{3.4}$$

$$SK_{x,y}^{i} = \sqrt[3]{\frac{1}{N^2} \sum_{j \in P} (O_i^j - M_{x,y}^{i})^3} \tag{3.5}$$

where, O_i^j is the O_i value of the jth pixel in P. The above three features are calculated for each of three color channels, yielding a nine-element texture feature vector:

$$V_{x,y}^{t} = [M_{x,y}^{1}, SD_{x,y}^{1}, SK_{x,y}^{1}, M_{x,y}^{2}, SD_{x,y}^{2}, SK_{x,y}^{2}, M_{x,y}^{3}, SD_{x,y}^{3}, SK_{x,y}^{3}] \tag{3.6}$$

Thus, by combining color and texture features, a 12-element feature vector for a pixel at (x, y) is obtained:

$$V_{x,y} = [V_{x,y}^{c}, V_{x,y}^{t}] \tag{3.7}$$

3.1.2.2 Object Classification

Once color and texture features were extracted for representing the characteristics of roadside objects including vegetation, the classification step builds a mapping from the features of each pixel in an input image to a class label that has the maximum probability across all classes. The task is accomplished using the widely used ANN, which accepts an feature vector $V_{x,y}$ for each pixel at (x, y) in an image as inputs, and outputs a probability value for each class:

$$P_{x,y}^{i} = tran(W_i V_{x,y} + b_i) \tag{3.8}$$

where, *tran* stands for the prediction function of a three-layer ANN which has a tan-sigmoid activation function in the hidden layer, and W_i and b_i are the trained weights and constant parameters for the ith object class. The pixel is assigned to the class label which has the highest probability across all classes:

$$C_{x,y} = \max_{i \in C} P^{i}_{x,y} \tag{3.9}$$

where, C stands for all object classes, and $C_{x,y}$ represents the class label for the pixel at coordinates (x, y) in an image.

3.1.3 Experimental Results

3.1.3.1 Evaluation Datasets and Metrics

The performance of the neural network approach is evaluated on the cropped roadside object dataset and a small set of natural roadside images. For the cropped roadside object dataset, we consider six types of objects (100 regions per object), including brown grass, green grass, tree leaf, tree stem, road and soil. There are a total of 600 cropped regions. The whole data is divided into two subsets (50 regions per object), which are used for two-fold cross validations in the experiments. A total of 10 natural roadside images were randomly selected from left-view DTMR video data. Note that no pixel-wise ground truths were created for these images.

Two performance measurement metrics are used: pixel-wise global accuracy measured in terms of all pixels across all testing images and all classes, and class accuracy averaged over each class using pixel-wise comparisons between classified results and ground truths. The global accuracy is biased favorably to frequently occurring object classes, and pays less attention to the low-frequency classes. By contrast, the class accuracy ignores occurring frequency of each class and treats the classification of all classes equally. Thus, they are able to reflect different aspects of the performance. The number of neurons in three layers of ANN is set to be 12–N–6, where N is the number of hidden neurons and determined experimentally. A resilient backpropagation algorithm is used for training the ANN.

3.1.3.2 Classification Results on Cropped Roadside Object Dataset

One key parameter of the ANN classifier is the number of hidden neurons used. To find an optimized number of hidden neurons, Fig. 3.2 shows the global accuracy versus the number of hidden neurons on both the training and test data, from which we can see that a larger number of hidden neurons is able to lead to constantly higher accuracy on the training data, while this is not the case for the test data where the accuracy fluctuates around 77% with the highest accuracy of 79% when 24 hidden neurons are used. The results imply the significance of designing a proper structure of ANN with optimized parameters to avoid overfitting on the training data and have stable performance on the test data. Thus, 24 hidden neurons are used in the neural network learning approach.

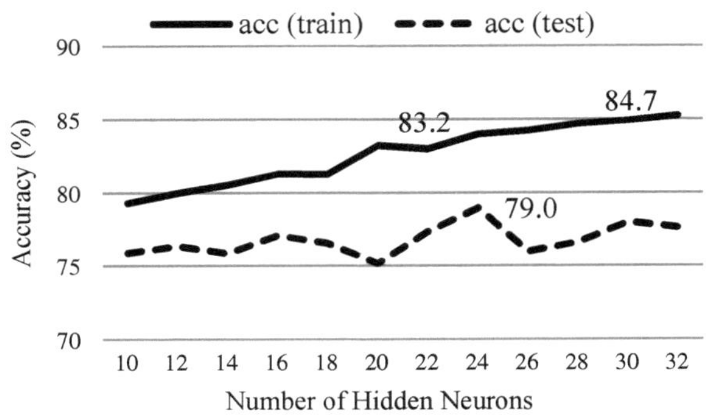

Fig. 3.2 Global accuracy versus number of hidden neurons in the ANN classifier

Table 3.1 presents the class accuracy on the test and training data using the ANN classifier with 24 hidden neurons. It can be seen that there are big differences in the accuracy between the training and test data for most of six classes. On the test data, tree leaf is the easiest class for correct classification, whereas brown grass is the most difficult one. By contrast, on the training data, road and tree stem become the easiest and most difficult classes respectively. The test data generally provides a more generalized and stable indication of the performance. It is worth noting that brown grass and soil are the two classes that are most prone to be misclassified to each other, which is probably due to their overlap in a yellow color, and thus more discriminative features may need to be added in order to further improve the accuracy for those objects. It is also noted that a significant proportion of tree stem pixels are misclassified as road, probably due to a similar black color and similar texture between them.

Table 3.2 compares the performance of the ANN classifier with commonly used SVM and nearest neighbor classifiers, and a combination of ANN, SVM, and nearest neighbor using a majority voting strategy. The implementation of SVM in the LIBSVM library [8] is adopted and two commonly used kernels—RBF and linear are compared. For the nearest neighbor classifier, the Euclidean distance is

Table 3.1 Class accuracy (%) on the test and training data

Data	Brown grass	Green grass	Road	Soil	Tree leaf	Tree stem
Test	65.9	88.3	84.4	68.4	94.0	68.9
Train	87.5	88.0	92.8	78.4	84.5	72.5

Table 3.2 Performance (%) comparisons between classifiers

Classifier	Train Acc.	Test Acc.
ANN	84.0	79.0
SVM (RBF)	99.7	44.0
SVM (linear)	77.7	75.5
Nearest Neighbor (NN)	100	68.6
Majority voting (linear SVM, ANN, NN)	87.2	76.1

used to calculate the dissimilarity (distance) between feature vectors. For fair comparison, the same set of color and texture features is used for all classifiers. It can be seen that the ANN performs the best on the test data, which is followed by a linear SVM, while the nearest neighbor ranks the last. A combination of three classifiers leads to higher accuracy than both the linear SVM and nearest neighbor, but still has lower accuracy than the ANN. It is interesting to observe that a RBF SVM demonstrates a typical overfitting problem in this case, as it achieves 99.7% accuracy on the training data, but the lowest accuracy of 44% on the test data. However, this is not true for the linear SVM, which implies the importance of choosing a suitable kernel for the SVM classifier on the evaluation data. The results indicate the performance of both ANN and SVM classifiers are heavily dependent on the parameters set for them.

Does the use of color intensity and moment fused features improve the classification performance, compared with using color features alone? To answer this question, we compare the performance of fused features versus color features using the same ANN classifier with 24 hidden neurons as shown in Table 3.3. Note that color features are comprised of three channels O_1, O_2, and O_3. From the table, we can see that the fused features seem to contain more useful and discriminative information about the objects, as they result in more than 11 and 6% higher accuracy respectively on the training and test data compared with using color features alone. The results confirm the benefit of adopting both color and texture information for vegetation classification in unconstrained real-world data.

3.1.3.3 Classification Results on Natural Images

We also qualitatively evaluate the neural network approach on a small set of roadside images that were randomly selected from left-view DTMR video data. Because there are no pixel-wise ground truths of object categories in these images, we analyze the performance of the approach by visually checking the correctness of the classification results.

Figure 3.3 shows six roadside images and their corresponding classification results. It can be observed that the major parts of tree, grass and soil are successfully detected using the neural network approach, which indicates the potential of using this approach in segmenting vegetation from other objects in practical applications. However, there are also some classification errors, which represent the typical challenges for vegetation classification in a realistic environment. For instance, the

Table 3.3 Performance (%) comparisons of color versus fused color and texture features based on the ANN classifier

Feature	Train Acc.	Test Acc.
Fused color and texture	84.0	79.0
Color	72.8	72.6

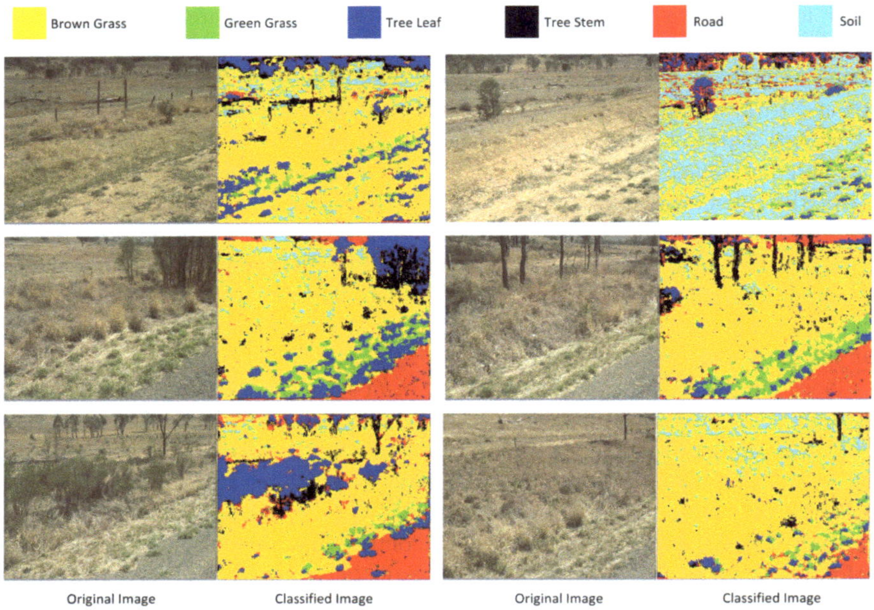

Fig. 3.3 Classification results of six sample images from the DTMR video data. One issue found is that pixels in object boundaries are prone to be misclassified, which is a common problem for patch based feature extraction techniques

shadow of trees is misclassified as tree stems in some images mainly because they share a similar black color, and a small proportion of green grasses are misclassified as tree leaves in some images, due to their overlap in a green color. The results seem to suggest that color information is insufficient for distinguishing all types of natural vegetation, particularly for those with a highly similar texture. Thus, employing more discriminative features to distinguishing trees from green grasses and handling shadows of objects are research directions that still need to be further investigated. Due to the use of patches for texture extraction, the boundary pixels between objects are also prone to be misclassified as road in those images. Thus, investigating a better solution to handling pixels in the boundaries between objects is another future direction in the field.

3.1.4 Summary

This section has described a neural network learning approach for object segmentation from natural roadside images. Color intensity and moment features are fused for more accurate classification with a multi-class ANN classifier. Experiments on the challenging cropped roadside object dataset indicate 79%

accuracy for classifying six objects, and encouraging results based on visual checks on a set of natural images. The results confirm that fusion of color and texture features leads to better performance than using color or texture features alone and it is still necessary to investigate more discriminative features to overcome challenges in natural conditions such as shadows of objects, objects with similar colors (e.g. green grass and tree leaf), and texture feature extraction at pixels in boundaries between objects.

3.2 Support Vector Machine Learning

3.2.1 Introduction

SVM is well-known for its capacity of handling complex non-linear problems and has been one of the most frequently used supervised learning algorithms in computer vision tasks. The basic concept of SVM is to construct a hyperplane (or a set of hyperplanes) in a multiple dimensional space that has the largest distance to the nearest training samples of any class. Learning the SVM classifier involves maximizing this distance, i.e. margin of the classifier, and the resulting hyperplane is fully specified by a subset of data samples, i.e. support vectors. Normally, non-linear feature samples are projected into a high-dimensional feature space by applying kernel functions so that a non-linear classification task is transferred to a more easily classified linear problem in the high-dimensional space. The kernel functions simply compute the inner products between pairs of data instead of directly operating the coordinates of data in high-dimensional feature spaces, and they require less computational time than the explicit computation of coordinates. Thus, a suitable kernel should be chosen in the use of an SVM classifier, and the most widely used ones include linear, polynomial, RBF and sigmoid. For more details about SVM, readers are referred to [9, 10].

In this section, we present an SVM learning approach [11] with a set of pixel characteristics features for segmenting grass, soil, road and tree regions from roadside video data. One of the main purposes is to automatically identify the proportion of a specific type of vegetation on the roadside.

3.2.2 SVM Learning Approach

The SVM learning approach [11] presents a pixel characteristics feature set for object representation and incorporates an SVM classifier for assigning each pixel to one of four object categories, including grass, soil, road and tree. Figure 3.4 describes the framework of the approach comprising of five main processing steps, including data acquisition which prepares data for the experiments, feature extraction which employs a set of pixel characteristics features for representing

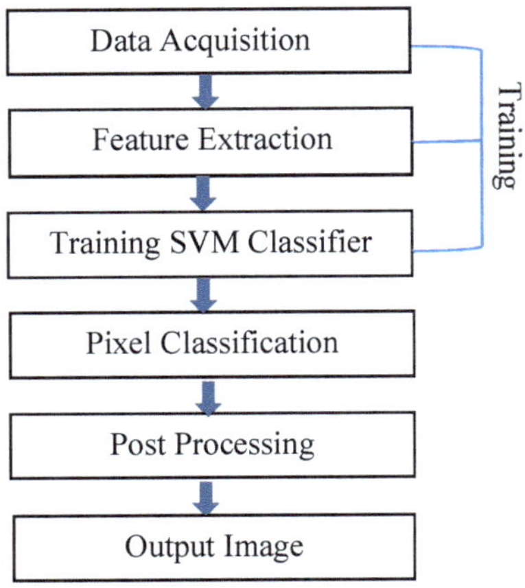

Fig. 3.4 Overview of the SVM learning approach

Fig. 3.5 (*Left*) Pixel values in grass regions; (*middle*) pixel values in soil regions; and (*right*) an input image with target regions labelled

roadside objects, classifier training which generates an SVM model based on the training data, pixel-based classification which labels each testing pixel to a class category, and post-processing which removes noise or misclassification error in the output images.

The SVM learning approach is similar to the approach [12] for terrain classification, which used a Gaussian mixture model based maximum likelihood classifier to discriminate between sand, grass and foliage. Different from [12], the SVM learning approach mainly utilizes pixel values in the RGB color space to discriminate different object categories. Figure 3.5 shows variations in pixel values in different roadside regions, and an image example with target regions labelled.

3.2.2.1 Color Feature Extraction

Given various types of color spaces available, the first step in extracting color features is to select a suitable space. For this purpose, we have conducted comparative tests on different types of color spaces. For each space, the extracted feature vectors are used to train the SVM classifier with a RBF kernel, which is further applied to classifying testing image pixels. We firstly ignored the YUV space because it is mainly used for television image enhancement. We started with the gray scale image, but failed to extract discriminative color information. Then we tried the CIELab space, and got some useful features but still did not obtain very good results. We further evaluated the YCbCr space and got quite promising results for segmenting vegetation and soil areas. However, under different environmental conditions, the results varied a lot and a large proportion of non-vegetation areas were falsely detected as vegetation. Then we moved to the RGB space and got the best performance. Therefore, we finally decided to generate color features based on R, G, and B color channels.

Table 3.4 lists the features in the pixel characteristics feature set that is used in the SVM learning approach. The feature set is composed of the absolute differences between pixel values of (R, G), (R, B) and (G, B), the average of R, G and B channels, the summation of R, G and B channels, and the summation of absolute differences between pixel values of (R, G), (R, B) and (G, B). The reason for taking absolute differences between pixel values lies in the fact that they need less calculation time and the differences between pixel values follow a similar proportion distribution which represents the same type of regions. The feature set is a main contribution in the SVM learning approach, as extracting an appropriate color feature set is still a challenge. The feature set is fed to the SVM classifier.

3.2.2.2 Training SVM Classifier

The SVM classifier is used to learn and distinguish different characteristics of roadside objects including grass, road, soil and tree. A key step in training the SVM classifier is to select a proper set of training data from different object regions. Using only green vegetation for the training makes the learnt SVM classifier suitable for identifying green vegetation, but not for yellow, red or brown vegetation. The training data includes 100 images of 1632×1248 pixels that were selected from the DTMR video data to represent all four objects.

Figure 3.6 shows the strategy used for pixel value acquisition from the training set with details of color features from a sample ROI. For an ROI of 5×5 pixels, we linearly get 25 pixel values. For each pixel, we define a feature set as shown in Fig. 3.6c. Each feature set belongs to one of four classes. The final training set is shown in Fig. 3.6d, where classes are defined in the first column and each class consists of multiple feature sets.

Table 3.4 A list of pixel characteristics features

No.	Feature
1–7	$\lvert R - G\rvert$, $\lvert R - B\rvert$, $\lvert G - B\rvert$, $(R + G + B)/3$, $2 * G - R - B$, $\lvert R - G\rvert + \lvert R - B\rvert + \lvert G - B\rvert$, $\lvert R + G + B\rvert$

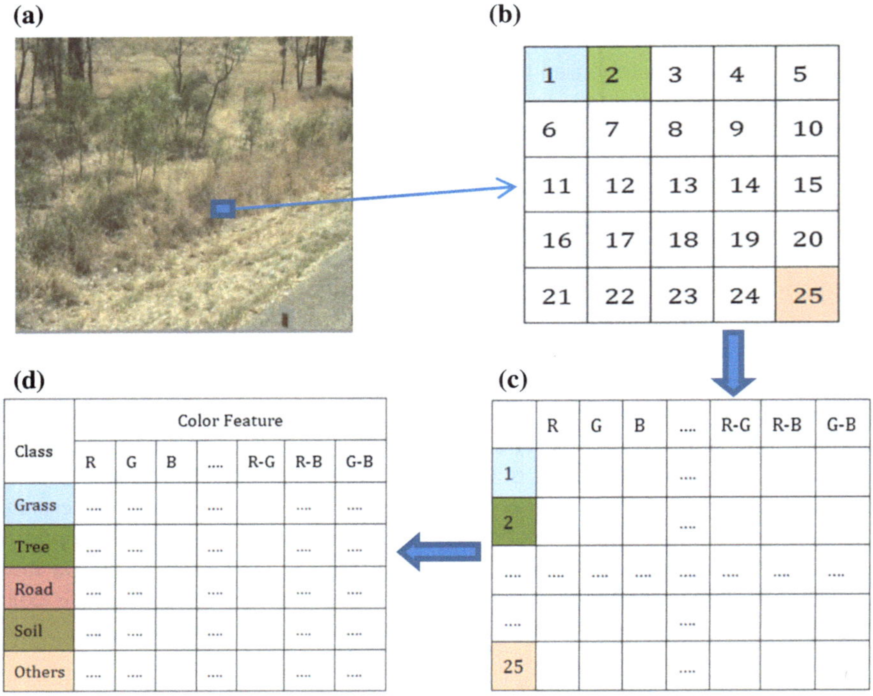

Fig. 3.6 Illustration of training data acquisition. **a** Original image with a ROI of 5 × 5 pixels; **b** index of pixels in the ROI; **c** the training data collected from the ROI; and **d** the final training data

Table 3.5 Average pixel value for each object

Average statistics	|R − G|	|R − B|	|G − B|	(R + G + B)/3
Brown soil	22	71	48	85
White soil	5	20	27	85
Green grass	20	17	30	85
Brown grass	13	40	27	85
Road	3	21	23	85
Tree	9	6	10	65

Table 3.5 shows the average pixel value for each class. The pixel values are randomly taken from the image data in each class. It can be seen that the average values of the first three statistics for different classes tend to be different, although the last parameter—average of R, G, and B is the same across all classes except for tree. The results indicate that the combination of these parameters is useful for classifying the objects, confirming the effectiveness of the extracted pixel characteristics feature set.

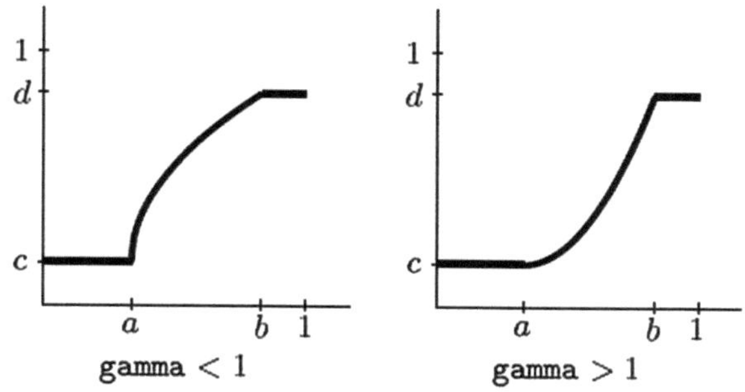

Fig. 3.7 Pixel value correction in the output images

3.2.2.3 Post-processing

In the classification results of a test image, each pixel is marked with a different color indicating different classes. To associate the pixel value with each class, some pixels are adjusted using Eq. (3.10). Figure 3.7 shows the adjustment of pixel values pictorially.

$$y = \left(\frac{x - a}{b - a}\right)^{\gamma}(\mathrm{d} - \mathrm{c}) + \mathrm{c} \tag{3.10}$$

It is observed from resulting images that some vegetation pixels are often inside a non-vegetation region and non-vegetation pixels inside a vegetation region. Those incorrectly identified regions should therefore be removed. One solution to moving these misclassified regions is based on the knowledge that soil regions are rarely found in the upper portion of the image where sky or tree regions often present. By removing those pixels incorrectly marked, we obtain regions of vegetation and soil.

3.2.3 Experimental Results

To evaluate the performance of the SVM learning approach, experiments with different types of roadside images collected at different roadside sites are conducted. Some input images are shown in Fig. 3.8. The performance of the approach is also compared to other existing approaches considering all environmental effects.

3.2.3.1 Qualitative Results

Figure 3.9 shows the results of region extraction from images, for grass, soil, and tree respectively. A yellow color is used to indicate grass regions, cyan for soil regions and black for tree regions. Overall, most of grass, tree and soil regions are

Fig. 3.8 Image samples used in experimental evaluations

(a) segmented grass pixels in a yellow color.

(b) segmented soil pixels in a cyan color.

(c) segmented tree pixels in a black color.

Fig. 3.9 Object segmentation results in sample images

accurately extracted from the images. The results show that the SVM learning approach is capable of segmenting the majority of pixels of different types of objects from roadside images, indicating that the pixel characteristics feature set works properly with an SVM classifier. However, there are also some mis-segmentation occurring in the output images. In Fig. 3.9a, a portion of road pixels are misclassified as grass and some shadow areas of grasses are incorrectly classified as road. A similar kind of misclassification also occurs for soil pixels, as some soil areas are not accurately identified while some grass pixels are misclassified as soil pixels. For small tree regions, they are not properly segmented due to low resolutions.

3.2.3.2 Quantitative Results

We used three measurements to quantitatively analyze the performance of the SVM learning approach, including: (1) a confusion matrix which is an effective way to present and analyze the performance of a classification model on the test data, (2) pixel-wise global classification accuracy which is calculated as the number of correctly classified samples divided by the total number of all test samples, and (3) precision which is calculated as true positive divided by the summation of true positive and false positive.

Table 3.6 shows the confusion matrix for four objects. The values indicate the number of pixels divided by a factor of 1000. Among the total 86,000 pixels, 73,000 pixels are correctly classified for tree, whereas for grass among the total 119,000 pixels, 90,000 pixels are correctly identified. 40,000 and 30,000 pixels are correctly classified for soil and road respectively. It is noted that there is a relatively low number of road pixels compared with other objects in the test images.

The overall accuracy of pixels achieved is 77%. Although this classification accuracy is relatively low, it is still acceptable since the focus is on the segmentation of the whole region rather than pixels. Because this accuracy is calculated based on pixels, many misclassified pixels are surrounded by correctly classified pixels, and those isolated pixels can be corrected by applying post-processing steps.

Table 3.7 displays the results of identifying grass regions based on the percentage of grasses in sample test images, showing high classification accuracy. To make a decision whether the target region is classified correctly or not, we set a threshold to the percentage difference for identifying vegetation regions. To be specific, if the difference between the ground truth and actually classified region is below or equal to 10%, the region is considered as classified correctly. Using this criterion, most of regions are properly identified, and we can decide whether a region has sparse, moderate or dense grasses.

Table 3.6 Confusion matrix for four objects

Actual	Predicted			
	Tree	Grass	Soil	Road
Tree	73	10	2	1
Grass	8	90	15	6
Soil	3	9	40	7
Road	0	2	5	30

Table 3.7 Classification of grass regions using percentages of grass pixels (%)

Image	Predicted	Actual	Diff.	Threshold	Decision
1	52	60	8	10	Classified
2	48	55	8	10	Classified
3	41	50	9	10	Classified
4	66	75	9	10	Classified
5	67	75	8	10	Classified
6	56	65	9	10	Classified
7	78	85	7	10	Classified
8	31	40	7	10	Classified
9	70	80	10	10	Classified
10	62	70	8	10	Classified

3.2.4 Summary

In this section, we have presented an SVM learning approach for the segmentation of grass, tree, road and soil regions from roadside images. A novel set of color features is extracted based on pixel color characteristics. The approach is evaluated on a set of realistic roadside images. Experimental results show that the approach distinguishes four regions with an overall pixel-level classification accuracy of 77% and reliably detects the percentage of grasses presented in images. The accuracy can be significantly increased once those isolated misclassified pixels are corrected based on prior knowledge of object distributions in images. As a future work, the approach still can be improved by including more pixel characteristics.

3.3 Clustering Learning

3.3.1 Introduction

Clustering learning is a commonly used unsupervised learning algorithm for classification problems. It generates a set of cluster centers, also called textons [13], to represent the characteristics of different objects. Its concept is similar to the bag-of-words representation model, and it constructs a vocabulary of visual words for an individual class or all classes via distance based clustering on filter response features from the training data, typically with simple but effective K-means clustering, and in the testing, each pixel is then assigned to the nearest neighboring texton, forming a frequency histogram of those textons which is taken as the image representation.

Many clustering learning approaches have been proposed previously for object segmentation and classification. For outdoor scene analysis, [14] constructed textons using K-means clustering from color and texture features comprising of L, a, b color channels and the difference in L between a pixel and its surrounding pixels.

A texton-based histogram was then built in a large neighborhood to merge similar clusters based on the Earth Movers Distance for segmenting objects. In [15], a universal visual vocabulary of texture textons was built by convolving the image with 17-D filter banks and aggregating the filter responses over all training images using K-means clustering. Extensions of textons include textonboost [16], which used 17-D filter bank textons in conjunction with color features to iteratively build a strong classifier via a sum of weak classifiers, and semantic texton forest [17], which combined a histogram of hierarchical semantic textons with a region's prior category distribution to build highly discriminative descriptors.

Most existing clustering learning approaches build generic textons for all classes from the training data and then map features into the closest texton, forming a histogram representation for the image. However, generic textons may not be effective to capture the specific characteristics of each class and handle the confusion between classes with similar characteristics. A histogram representation may fail for small images due to a sparse bin problem. Texton features have seldom been used for vegetation segmentation and classification previously.

This section presents a clustering learning approach [18] which utilizes superpixel based class-semantic texton occurrences for natural roadside object segmentation. A similar pervious work is [19], which built a set of semantic SIFT words for each object and then integrated all words over the image for scene understanding. The visual words were formed by performing K-means clustering on SIFT features extracted from local image bocks. The SIFT features in test images were mapped to the semantic visual words and object classification was accomplished by performing majority voting on the number of mapped visual words in an image over all classes. But the approach did not consider color textons and assumed each cropped testing image belonging to only one object. The evaluation was restricted to only a few manually cropped images. These issues are addressed in the clustering learning approach.

3.3.2 *Clustering Learning Approach*

3.3.2.1 Approach Framework

Figure 3.10 depicts the framework of the clustering learning approach, which comprises of a training stage and a testing stage. During the training stage, an equal set of local regions is manually cropped from the training data for each class. Color and filter bank responses are extracted from those regions, and further fed into K-means clustering to create two individual sets of class-semantic color and texture textons for each class. Each set of color or texture textons is combined for all classes to form two class-semantic texton matrices—one for color and the other for texture. At the testing stage, the input image is firstly segmented into a set of heterogeneous superpixels, and color and filter bank features of all pixels in each superpixel are then extracted and projected separately into one of the learnt color or

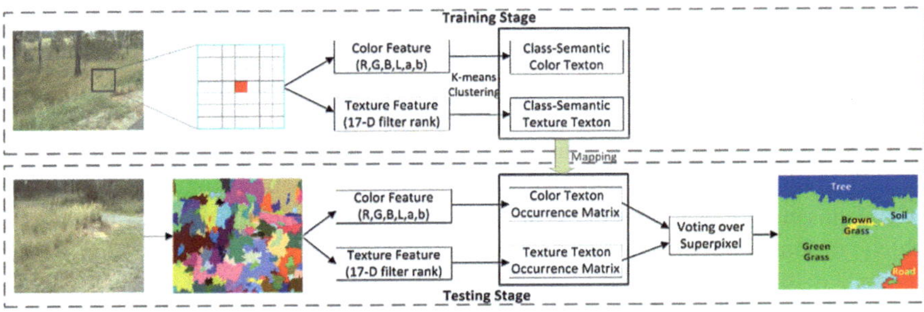

Fig. 3.10 Framework of the clustering learning approach. During training, a set of regions is cropped to generate class-semantic color and texture textons using K-means clustering. During testing, features of all pixels within each superpixel are projected to one of the learnt textons, and further aggregated into texton occurrences. A majority vote over texton occurrences in each superpixel is used to segment objects from roadside images

texture textons using the Euclidean distance. The superpixel based color and texure texton occurrences for each class can then be obtained and further combined using a linear mixing method. Finally, object segmentation is achieved by assigning all pixels in each superpixel to a class label which has the maximum combined occurrence across all classes, including brown grass, green grass, soil, road, tree leaf, tree stem, and sky.

3.3.2.2 Superpixel Acquisition

The first pro-processing step of the clustering learning approach is to segment an input test image into a set of local superpixels which have homogenous appearance and each of them is anticipated to belong to only one class category. The approach takes superpixels as the basic processing unit due to their compelling characteristics in simplifying the classification problem and utilizing collective decisions over a pool of pixels within each superpixel. Thus superpixel based classification is expected to be able to significantly reduce the complexity of the categorization process in the clustering learning approach. Amongst many popular region segmentation algorithms, such as mean shift [20], JSEG [21] and superpixel [22], we employ the graph-based algorithm [23] due to its fast processing speed and high reported performance in natural scene analysis.

3.3.2.3 Color and Texture Feature Extraction

Feature extraction aims to extract a set of discriminative visual features to distinguish the appearance of different object categories. The generation of textons in the clustering learning approach is based on two types of features: color and texture, which are expected to be able to complement each other for a more effective feature representation of roadside objects.

Color: one criterion in choosing a suitable color space is that the color space should be perceptually consistent with human vision, as the human eye is very adept about distinguishing one object from others even under extremely challenging environmental conditions. The clustering learning approach adopts the CIELab space, which has high perceptual consistency with human vision and demonstrated a good generalized performance on scene content understanding [17]. We also include RGB as it may contain complementary information that is vital for recognizing specific objects. The color feature vector of a pixel at the coordinate (x, y) in an image is composed of:

$$\mathrm{V}^{\mathrm{c}}_{\mathrm{x,y}} = [R, G, B, L, a, b] \tag{3.11}$$

Texture: there are a wide range of filters that have been proposed previously for generating textons for object segmentation, such as Leung and Malik with 48 filters, Schmid set with 13 filters, maximum response (MR8) set with 38 filters, and Gabor set with a certain number of filters, etc. The clustering learning approach adopts 17-D filter banks which were firstly adopted in [15], and have shown high performance for generic object classification. The 17-D filter banks include Gaussians with 3 different scales (1, 2, 4) applied to L, a, and b channels, Lapacians of Gaussians with 4 different scales (1, 2, 4, 8), and the derivatives of Gaussians with 2 different scales (2, 4) for each axis (x and y) on the L channel. By convolving each image with the filter banks, 17 response images can be obtained and each pixel is characterized by 17 responses. For a pixel at (x, y) in an image, its texture feature vector is composed of:

$$\mathrm{V}^{\mathrm{t}}_{\mathrm{x,y}} = [\mathrm{G}^{\mathrm{L}}_{1,2,4}, \mathrm{G}^{\mathrm{a}}_{1,2,4}, \mathrm{G}^{\mathrm{b}}_{1,2,4}, \mathrm{LOG}^{\mathrm{L}}_{1,2,4,8}, \mathrm{DOG}^{\mathrm{L}}_{2,4,\mathrm{x}} \mathrm{DOG}^{\mathrm{L}}_{2,4,\mathrm{y}}] \tag{3.12}$$

3.3.2.4 Class-Semantic Color-Texture Texton Construction

After obtaining the color and texture features, we proceed to generate two individual sets of middle-level texton features from each of them. Unlike existing texton based approaches that generate a universal visual vocabulary for all classes, the clustering learning approach extracts a representative set of the most discriminative textons specifically for each class, i.e. class-semantic textons, which are expected to represent more separable and less redundant characteristics for each object to reduce confusion between classes.

Let assume there are C classes and n training pixels in the ith class (i = 1, 2,…, C), and V_i^c and V_i^t be the color and texture feature vectors respectively for the ith class, the K-means clustering algorithm is employed to generate a set of textons for each of V_i^c and V_i^t by minimizing:

$$J_c = \sum_{j=1}^{n} \min_k \left| V_{i,j}^c - T_{i,k}^c \right|^2 \tag{3.13}$$

where, $V_{i,j}^c$ is the color features of the jth pixel in V_i^c, $T_{i,k}^c$ is the kth color textons learnt (k = 1,2,…,K) for the ith class, and J_c is the error function. The function for texture features is similar to (3.13). The ith class-semantic color and texture texton vectors are respectively composed of:

$$T_i^c = [T_{i,1}^c, T_{i,2}^c, \ldots, T_{i,K}^c] \tag{3.14}$$

$$T_i^t = [T_{i,1}^t, T_{i,2}^t, \ldots, T_{i,K}^t] \tag{3.15}$$

The textons are basically the cluster centres of color or texture features that have the minimum Euclidean distance between them and all feature descriptors. The value of K controls the number of learnt textons and determines the size of the texton feature space, which often has significant impact on the effectiveness of the learnt textons in representing the characteristics of each class.

Combine the color or texture texton vector for all C classes, a color and a texture texton matrix can be formed respectively:

$$T^c = \left\{ \begin{matrix} T_{1,1}^c, T_{1,2}^c, \ldots, T_{1,K}^c \\ T_{2,1}^c, T_{2,2}^c, \ldots, T_{2,K}^c \\ \vdots \\ T_{C,1}^c, T_{C,2}^c, \ldots, T_{C,K}^c \end{matrix} \right\} \quad \text{and} \quad T^t = \left\{ \begin{matrix} T_{1,1}^t, T_{1,2}^t, \ldots, T_{1,K}^t \\ T_{2,1}^t, T_{2,2}^t, \ldots, T_{2,K}^t \\ \vdots \\ T_{C,1}^t, T_{C,2}^t, \ldots, T_{C,K}^t \end{matrix} \right\} \tag{3.16}$$

The above two matrices are comprised of color and texture textons for all C classes learnt from the training data respectively, and are expected to contain representative and discriminative features for each class, which are used for distinguishing between objects in the test data.

3.3.2.5 Superpixel Based Texton Occurrence and Object Segmentation

For all pixels P^I in an image and a set of object categories C^N, the task of object segmentation is to find a mapping function $M : P^I \rightarrow C^N$ so that each pixel corresponds to a category. Given the learnt color and texture texton matrices, the clustering learning approach adopts a majority voting classification strategy to obtain the class label for all pixels in a test image based on superpixel based texton occurrences, which essentially aggregates texton occurrences over all pixels in each superpixel to reach a collective classification decision. To be specific, we first map all pixels in the test image into one of the learnt color textons and one of the learnt texture textons respectively, and then calculate the occurrences of mapped color and texture textons of all pixels in each superpixel for each class. The color and texture texton occurrences are further combined using a linear mixing method to obtain the class probabilities for each superpixel, which indicate the likelihood of this superpixel belonging to each of all classes, and all pixels in this superpixel are finally assigned to the class label with the highest probability across all classes.

For an input image I, it is firstly segmented into a set of superpixels with homogenous features using a fast graph-based algorithm [23]:

$$S = [S_1, S_2, \ldots, S_L] \tag{3.17}$$

where, L is the number of segmented superpixels, and S_l indicates the lth superpixel.

Let assume there are m pixels in the superpixel S_l, the color and texture feature vectors of S_l can be extracted using Eqs. (3.12) and (3.13) respectively, i.e. $V_{S_l}^c = \bigcup_{x,y \in S_l} V_{x,y}^c$ and $V_{S_l}^t = \bigcup_{x,y \in S_l} V_{x,y}^t$. The $V_{S_l}^c$ and $V_{S_l}^t$ are projected to the learnt class-semantic color and texture textons respectively by searching the closest texton using the Euclidean distance metric:

$$\mathrm{f}\left(V_{x,y}^c, T_{i,k}^c\right) = \begin{cases} 1, & \mathit{if} \left\| V_{x,y}^c - T_{i,k}^c \right\| = \min_{q=1,2,\ldots C; p=1,2,\ldots,K} \left\| V_{x,y}^c - T_{q,p}^c \right\| \\ 0, & \mathit{otherwise} \end{cases} \tag{3.18}$$

A color texton occurrence matrix which records the number of the mapped textons $T_{i,k}^c$ for all pixels in S_l can be obtained using:

$$A_{i,k}^c(S_l) = \sum_{x,y \in S_l} \mathrm{f}(V_{x,y}^c, T_{i,k}^c) \tag{3.19}$$

The figures in the color texton occurrence matrix are then accumulated for the ith class, yielding the occurrence of color textons in the superipixel S_l for this class:

$$A_i^c(S_l) = \sum_{k=1}^{K} A_{i,k}^c(S_l) \tag{3.20}$$

Repeat the above procedure for the texture feature vector $V_{S_l}^t$ to obtain the occurrence of texture textons in S_l for the ith class:

$$A_i^t(S_l) = \sum_{k=1}^{K} A_{i,k}^t(S_l) \tag{3.21}$$

Combine the occurrences of color and texture textons using a simple linear mixing method to generate a combined occurrence in S_l for the ith class:

$$A_i^l = A_i^c(S_l) + w \times A_i^t(S_l) \tag{3.22}$$

where, w is a weight for texture textons relative to a fixed value of 1 for color textons, and it indicates the relative contribution of texture textons to the combined results. The combined occurrence is further converted into a class probability by dividing the total number (i.e. M) of all pixels within S_l:

$$p_i^l = A_i^l / M \tag{3.23}$$

A class probability vector of S_l for all classes can be obtained:

$$P^l = [p_1^l, p_2^l, \ldots, p_C^l] \tag{3.24}$$

All pixels in S_l are finally assigned to the cth class with the maximum class probability over all classes:

$$S_l \in c\text{th class if } p_c^l = \max_{i=1,2,\ldots C} p_i^l \tag{3.25}$$

The above procedure makes a collective classification decision for each superpixel based on color and texture texton occurrences of all pixels within the superpixel, to utilize supportive information in a spatial neighbourhood. Thus, the result is expected to be robust to small error or noise in the superpixel. Note that the pre-processing step of image segmentation is not performed for data in the cropped roadside object dataset, which have only one object in each, and each of them is treated as an individual superpixel.

3.3.3 Experimental Results

The performance of the clustering learning approach is evaluated on the cropped roadside object dataset and the natural roadside object dataset. Different values of several key parameters are evaluated comparatively to achieve a balanced performance between accuracy and computational time, and we further apply the approach to the practice of object segmentation on real-world roadside video data.

3.3.3.1 Implementation Details and Parameter Settings

All natural images are scaled to a fixed size of 320 × 240 pixels to facilitate the processing of image segmentation and reduce computational costs. The parameters of the graph-based image segmentation algorithm are the same as the recommended settings in [24], i.e. $\sigma = 0.5$, $k = 80$, and $min = 80$ for an image size of 320 × 240 pixels. To ensure balanced training data between objects, 120 pixels are selected at random coordinates in each cropped region and used for generating the color-texture textons using K-means clustering. The number of color and texture textons is set to be the same in the fusion of color and texture textons, i.e. color-texture textons, on both the cropped and natural roadside object datasets. The whole system was implemented under a Matlab platform using a Macbook laptop with a configuration of 1.8 GHz Intel Core i5 processor and 4 GB memory.

Evaluation metrics: The performance of the clustering learning approach is evaluated using two measurements: global accuracy and class accuracy. Four cross-validations are used to obtain an average accuracy. In detail, the cropped regions for each class are split into four equivalent subsets, and for each cross-validation, three subsets are used for training and the left one for testing.

3.3.3.2 Global Accuracy Versus Number of Textons

Figure 3.11 shows the global accuracy of the clustering learning approach versus the number of textons in classifying seven objects on the cropped roadside object dataset, and classifying six objects on the natural roadside object dataset. Three

Fig. 3.11 Global accuracy versus the number of textons. The number of color and texture textons is the same, the combination weight for texture textons $w = 1$, the size of Gaussian filters: 7×7 pixels, and distance metric: Euclidean

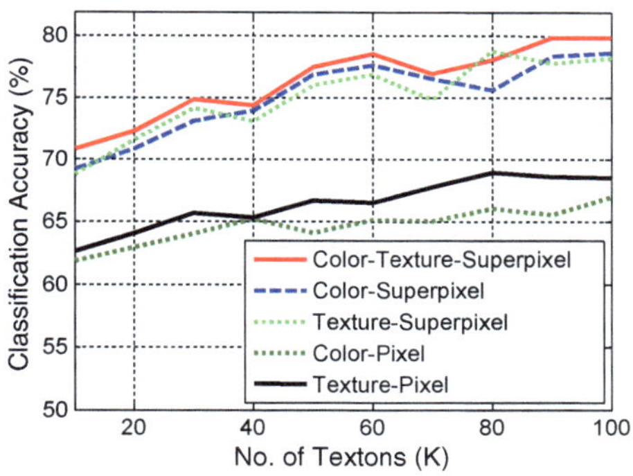

(a) cropped roadside object dataset

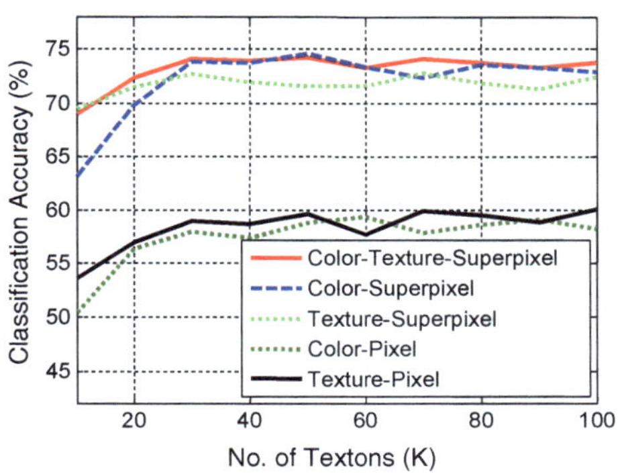

(b) natural roadside object dataset

types of features are compared, including color-texture textons, color textons alone, and texture textons alone, as well as two types of classification strategies: superpixel based collective decision (i.e. superpixel) and pixel based single decision (i.e. pixel). Note that color-texture textons represent a combination of equal sets of color and texture textons.

We can observe that superpixel based classification has much higher accuracy (around 14%) than pixel based classification for both the cases of using color or texture textons on the two datasets. This proves the benefit of aggregating collective classification decisions over a pool of pixels within each superpixel, which leads to more accurate results than pixel based classification. For both superpixel and pixel based classification and both datasets, color textons and texture textons exhibit a similar overall performance. Their performance increases gradually with a larger number of textons on the cropped dataset, but tends to level off on the natural image dataset. For both datasets, a combination of color and texture textons (i.e. color-texture textons) leads to slightly higher global accuracy than using color or texture textons alone. The highest global accuracies of 79.9 and 74.2% are obtained using 90 color-texture textons on the cropped dataset, and using 30 color-texture textons on the natural image dataset, respectively. Compared with using color or texture textons alone, using color-texture textons obtains higher global accuracy with less textons, which is important for applications that require real-time processing.

Figure 3.12 reveals the computational time versus the number of color-texture textons on the two datasets. The overall computation is the average seconds required for each test image (or region), which is primarily comprised of two processing stages: feature extraction and texton classification. For both datasets, there is an approximately linear relationship between the overall computation and the number of textons, and the vast majority of the overall time is used for texton mapping and classification. By contrast, the time used for color and texture feature extraction remains constant and takes only a small proportion of the overall time. Due to higher resolutions, natural images require more time than cropped regions. Thus, it is important to choose a proper number of color-texture textons to achieve a

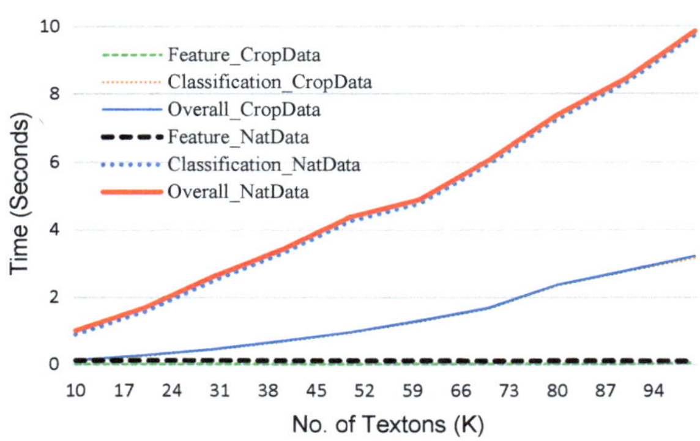

Fig. 3.12 Computational performance versus the number of textons on the cropped and natural roadside object datasets. The time is the average seconds per image (or region) required for calculating color and texture features (i.e. Feature), and performing texton mapping and classification (i.e. Classification)

good balance between accuracy and computation. The clustering learning approach chooses to use 60 and 30 color-texture textons for the cropped and natural roadside object datasets respectively, for which the accuracies are 78.6 and 74.2%, and the computational time is 1.3 and 2.6 s respectively.

3.3.3.3 Global Accuracy Versus Combination Weights

We examine the impact of the weight given to texture textons in their combination with color textons on the global accuracy. The weight is a value relative to a fixed value of 1 for color textons, and it indicates the contribution of texture textons to the combined results. Figure 3.13 shows the results of the clustering learning approach in terms of global accuracy using a *w* value ranging from [0.1, 1.5], and the results are obtained using 60 and 30 color-texture textons on the cropped and natural roadside object datasets, respectively. For both datasets, fusion of color and texture textons has higher global accuracy than using color or texture textons alone for most weight values, and the best overall performance is obtained using weight

Fig. 3.13 Global accuracy versus the value of combination weights. The weight of texture textons is a value relative to a fixed value of 1 for color textons. Color & texture texton no. = 60 (cropped dataset) and 30 (natural dataset); the size of Gaussian filters: 7 × 7 pixels; distance metric: Euclidean

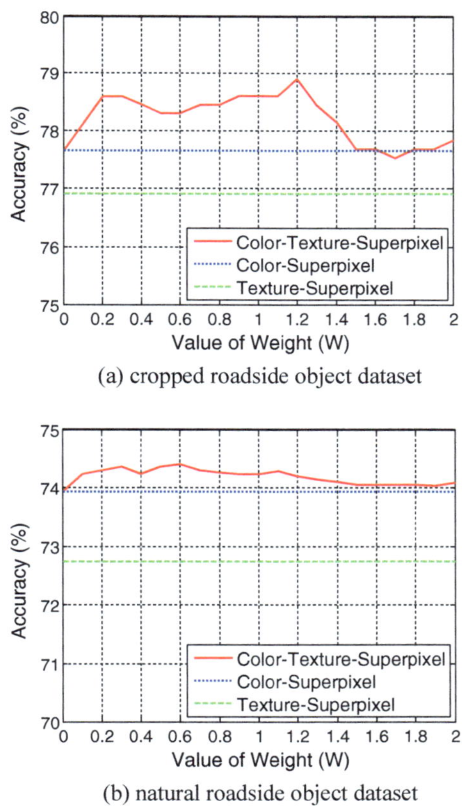

values between 0.2 and 1.2. This indicates that color textons play a slightly more significant role than texture textons in the classification of roadside objects using the clustering learning approach. The highest accuracy achieved is 78.9% for the cropped dataset when w equals to 1.2 and 74.4% for the natural dataset when w equals to 0.6.

3.3.3.4 Global Accuracy Versus Size of Gaussian Filters

We also investigate the impact of the size of Gaussian filters on the global accuracy as shown in Fig. 3.14. The size determines the range of a spatial neighborhood from which texture features are extracted using Gaussian filters, and thus they may have substantial impact on the effectiveness of the generated texture textons in representing discriminative characteristics of each object. The performance of three approaches—superpixel based classification using color-texture textons, pixel or superpixel based classification using texture textons, are compared using five sizes of Gaussian filters, ranging from 5×5 to 15×15 pixels with an interval of 2 pixels. The results are obtained using 60 and 30 color-texture textons on the cropped and natural datasets, respectively. For all the three methods and both datasets, there are only small differences in the accuracy using five sizes of Gaussian filters, but using small sizes seems to slightly outperform using high sizes, particularly for the cropped dataset. The highest accuracies of 78.9 and 74.6% are achieved using sizes of 7×7 and 9×9 respectively for the two datasets. Considering the fact there is only a small performance difference between using sizes of 7×7 and 9×9 on the natural roadside object dataset, a size of 7×7 is used for both datasets in the clustering learning approach.

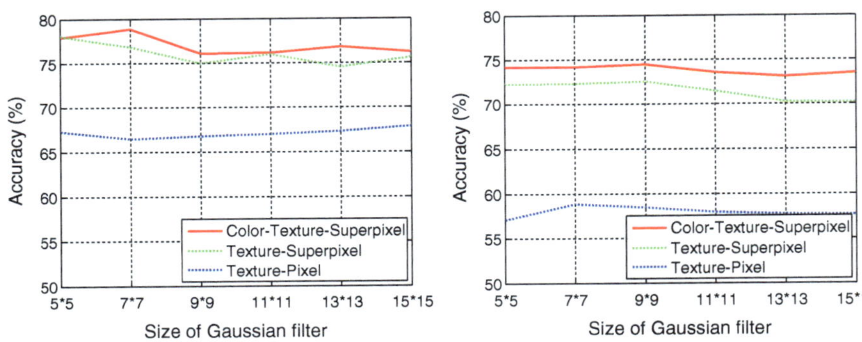

(a) cropped roadside object dataset (b) natural roadside object dataset

Fig. 3.14 Global accuracy versus the size of Gaussian filters. The combination weight $w = 1.2$ (cropped dataset), and 1 (natural dataset); Color and texture texton no. = 60 (cropped dataset) and 30 (natural dataset); distance metric: Euclidean (for both datasets). The performance of color textons remains the same using varied sizes of Gaussian filters

3.3.3.5 Global Accuracy Versus Distance Metrics

Another important factor impacting the generation of textons is the distance metric used in K-means clustering. The metric determines the measurement unit that is used to calculate the dissimilarity between color or texture features. Table 3.8 compares the results of four distance metrics—Squared Euclidean, Sum of absolute differences (CityBlock), Cosine, and Correlation. We can observe that the use of different distances exerts only small influence on the performance of using color-texture textons or using color textons, but large influence on that of using texture textons. Among four metrics, the Euclidean has the highest accuracy for both color-texture textons and texture textons, whereas the Cityblock produces the highest accuracy for texture textons on the two datasets. We investigate the results of color-texture textons when the Euclidean is used for generating color textons and the Cityblock for generating texture textons. Our results show 77.5% accuracy on the cropped roadside object dataset, which is lower than the results of using the Euclidean for both color and texture textons.

3.3.3.6 Class Accuracy and Confusion Analysis

Table 3.9 compares class accuracy of the clustering learning approach to five other approaches. It can be seen that color-texture textons have the highest global accuracy on both datasets, and the highest average class accuracy on the cropped dataset. Supervised color textons perform better in classifying road pixels than texture textons, while texture textons are better in classifying soil and sky pixels on the cropped dataset. This result, however, is not observed on the natural roadside object dataset where color textons show higher class accuracies than texture textons for all objects, except for brown grass. Compared to superpixel based classification, there are significant reductions on the class accuracy for all objects on both datasets using pixel based classification, particularly for brown grass and road which have nearly 20% reductions. This indicates the importance of utilizing a superpixel based collective decision rather than a pixel based single decision for object classification.

Table 3.10 displays the confusion matrix of objects using color-texture textons. For both datasets, sky is the easiest object for correct classification with more than 96% accuracy, and road is also classified with high accuracy. The results agree with

Table 3.8 Global accuracy (% ± standard deviation) versus distance metrics

Dataset	Texton	Euclidean	CityBlock	Cosine	Correlation
Cropped	Color-texture	78.9 ± 3.4	75.5 ± 2.2	78.0 ± 5.0	78.1 ± 2.2
	Color	77.7 ± 2.3	73.5 ± 3.3	76.8 ± 3.5	77.5 ± 3.2
	Texture	76.9 ± 4.4	78.1 ± 3.8	71.9 ± 5.3	72.5 ± 3.7
Natural	Color-texture	74.2	74.2	72.1	72.3
	Color	73.9	72.9	71.8	72.0
	Texture	72.8	73.3	65.1	66.6

Table 3.9 Comparisons of class accuracy (%) between approaches

(a) Cropped roadside object dataset

		Brown Grass	Green Grass	Road	Soil	Tree leaf	Tree stem	Sky	Avg.	Global
Superpixel based	Color-texture	86.2	88.0	84.2	57.0	79.0	68.0	96.1	79.8	78.9
	Color	85.2	87.0	90.1	51.0	80.0	65.0	92.2	78.6	77.6
	Texture	85.2	88.0	80.2	59.0	76.0	62.0	98.0	78.3	76.9
Pixel based	Color	56.0	65.3	66.5	48.7	58.5	56.6	87.8	62.8	65.0
	Texture	52.4	71.3	68.0	44.4	55.3	55.1	91.4	62.6	67.2

(b) Natural roadside object dataset

		Brown grass	Green grass	Road	Soil	Tree	Sky	Avg.	Global
Superpixel based	Color-texture	73.5	78.7	85.7	42.5	67.6	96.6	74.1	74.2
	Color	71.3	81.7	83.9	44.4	67.9	98.1	74.6	73.9
	Texture	74.2	73.7	83.7	36.3	65.5	94.1	71.3	72.8
Pixel based	Color	47.4	65.1	66.9	41.0	61.7	94.0	62.7	58.0
	Texture	54.9	56.4	70.1	35.6	59.4	88.4	60.8	59.0

Table 3.10 Confusion matrix (%) for different classes using the clustering learning approach

	Brown grass	Green grass	Road	Soil	Tree leaf	Tree stem	Sky
(a) Cropped roadside object dataset							
Brown Grass	**86.2**	0	1.0	6.9	0	5.9	0
Brown Grass	1.0	**88.0**	0	0	11.0	0	0
Road	0	0	**84.2**	1.0	0	11.8	3.0
Soil	33.0	0	3.0	**57.0**	0	4.0	3.0
Tree Leaf	1.0	11.0	1.0	0	**79.0**	8.0	0
Tree Leaf	8.0	0	17.0	4.0	3.0	**68.0**	0
Sky	0	0	1.9	2.0	0	0	**96.1**
(b) Natural roadside object dataset							
	Brown grass	Green grass	Road	Soil	Tree	Sky	
Brown Grass	**73.5**	14.7	3.3	3.1	5.4	0.0	
Brown Grass	7.8	**78.7**	2.4	0.5	8.6	0.0	
Road	7.3	0.4	**85.7**	6.1	0.1	0.4	
Soil	39.6	5.8	7.0	**42.5**	5.1	0.0	
Tree	5.8	5.4	18.0	0.3	**67.6**	2.9	
Sky	0.2	0.0	2.8	0.1	0.3	**96.6**	

The bold figure indicates class accuracy for each object

previous results [25, 26], where sky and road have the highest classification accuracies among five objects on the OU and MA datasets, and among eight objects on a road scene video dataset, respectively. By contrast, soil is the most difficult class with only 57.0 and 42.5% accuracies respectively on the two datasets, and a significant proportion (more than 33%) of soil pixels are misclassified as brown grass, probably due to the overlap in a yellow color. Furthermore, more than 17% of tree pixels are misclassified as road. A similar result was also observed in [25] that the top part of some tree leaves was wrongly recognized as road. The results imply the necessity of adopting more discriminative texture features that are specifically designed to distinguish between them under natural conditions. Unlike the result on the cropped dataset that there is little confusion between brown and green grasses, brown and green grass pixels are prone to be misclassified to each other on the natural roadside object dataset, indicating typical challenges for robust vegetation segmentation in natural images that are not observed in manually cropped regions.

Figure 3.15 displays samples of segmentation results, showing promising overall accuracy. The results on these samples visually confirm the confusion results between objects in Table 3.10. The confusion between brown and green grasses is partially due to the difficulty of manually creating accurate ground truths for them. Tree pixels are prone to be misclassified as road, due to their similarity in

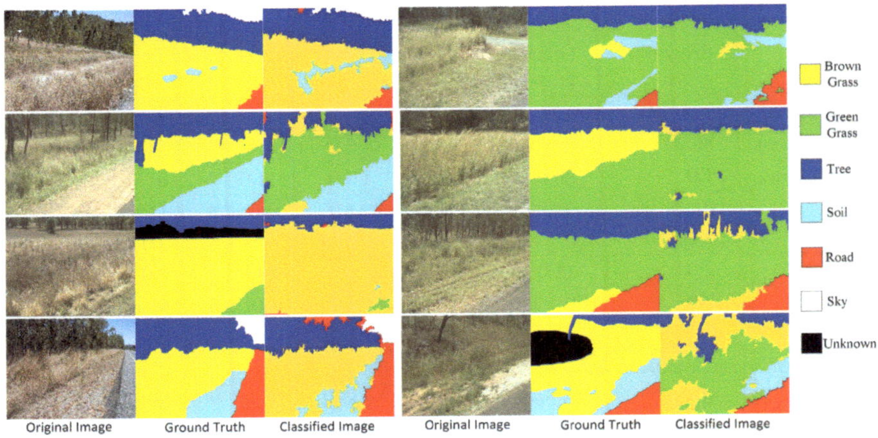

Fig. 3.15 Samples of segmentation results on the natural roadside object dataset

texture and overlap in a dark green color. Similarly, soil and brown grasses also tend to be misclassified to each other, due to a similar yellow color.

3.3.3.7 Application to Natural Roadside Video

We also apply the clustering learning approach to the practice of vegetation segmentation on a set of 36 videos taken by the DTMR on two state roads within the Fitzroy region, Queensland, Australia. Figure 3.16 shows a subset of original frames from three sample videos and their corresponding segmentation results. These frames are manually selected to be representative of different scene content and varying environmental conditions so that they can reveal the performance of the clustering learning approach under realistic scenarios. It can be seen from the figure that the majority of grass and tree regions are classified successfully, proving the effectiveness and direct applicability of the clustering learning approach to supporting real-world applications. The results also illustrate a small portion of misclassified pixels, which reveals typical challenges for vegetation segmentation on real-world video data. To be specific, regions accompanied by shadows of objects are prone to be misclassified as tree stems, primarily due to their similar characteristics of a dark color. For the same reason, a small proportion of road regions is also misclassified as tree stems. A similar effect of shadows was also observed in [25], where the presence of shadows leads to misclassification of trees to unknown objects in outdoor scenes. This indicates the importance of handling the impact of lighting variations for more accurate segmentation on natural data. In addition, there also exist some confusion between soil and brown grass pixels, due to their similarity in a yellow color particularly under bright lighting conditions. The results

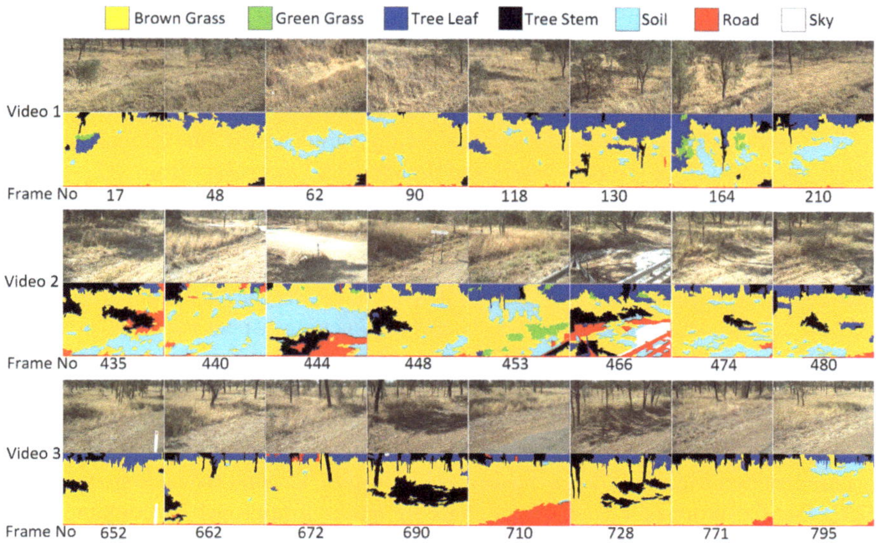

Fig. 3.16 Segmentation results on frame samples from three roadside videos. Frames in the three rows are selected from the beginning, middle, and end sections of the video, respectively. The frame number in each video is shown below each figure

reflect the fact that color is still the dominant impacting factor in leading to the confusion between objects, and thus it is still necessary to incorporate more effective texture features to further improve the results.

3.3.3.8 Performance Comparisons

For performance comparisons with deep learning approaches, we firstly report the performance of the popular CNNs on the cropped roadside object dataset as shown in Table 3.11. Due to low image resolutions of cropped regions in the dataset, the LeNet-5 [27], which was initially designed for handwritten recognition, is used here. Note that recently developed CNN models such as AlexNet, VGG-19 and GoogLeNet cannot be directly used here as they were built for larger image resolutions, e.g. 224×224 pixels. To keep a consistent size of input data, all cropped images are resized into $W \times W$ pixels $W \in \{32, 64, 128\}$ using (1) a varied ratio method which resizes both the width and height of regions into W pixels and does not keep the aspect ratio of the width to the height, and (2) a fixed ratio method

Table 3.11 Global accuracy (%) of the LeNet-5 CNN on the cropped roadside object dataset

Image size (pixels)	32 × 32	64 × 64	128 × 128
Varied ratio of width to height	73.5	67.9	55.6
Fixed ratio of width to height	64.8	75.9	34.0

which resizes the larger one between the width and height of regions into W pixels and keeps the aspect ratio. Studies [28] indicated that keeping the aspect ratio of an image helps preserve the shape of objects and boost the performance. The evaluations are based on four random cross-validations.

We can see that, for regions resized using varied ratios, the global accuracy drops from 73.5 to 55.6% when the image size increases from 32 × 32 to 128 × 128 pixels, and this is probably because using a small size helps prevent substantial information loss due to resizing small regions into a larger size. When fixed ratios are used, using 64 × 64 pixels produces the highest accuracy of 75.9%, and this is expected as image resizing is based on the larger value of the width and height. The results also confirm the benefit of keeping aspect ratios of images in minimizing object distortion and achieving higher accuracy. By contrast, using 128 × 128 pixels has the lowest accuracy for both varied and fixed ratio methods, due to low resolutions of cropped regions.

Table 3.12 compares the clustering learning approach with existing approaches on the cropped roadside object dataset and the public Croatia roadside grass dataset [29]. For the cropped dataset, the clustering learning approach is compared with four types of approaches: (1) the generic texton histogram approach which creates a universal set of textons for all classes using K-means clustering, and classifies every cropped region to a class which has the nearest distance of texton histogram with

Table 3.12 Performance comparisons with existing approaches

Dataset	Approach	Classifier	Object No.	Resolution	Acc. (%)
Cropped	Color-texture texton (proposed)	KNN	7	–	78.9
	Generic texton histogram	KNN	7	–	59.7
	LeNet-5 [27]	–	7	64 × 64	75.9
	Color statistic [11]	SVM	7	–	77.0
	$O_1O_2O_3$ + color moment [3]	ANN	6	–	79.0
	$O_1O_2O_3$ + color moment [3]	SVM	6	–	75.5
	$O_1O_2O_3$ + color moment [3]	KNN	6	–	68.6
	$O_1O_2O_3$ [3]	ANN	6	–	72.6
Croatia	Color-texture texton (proposed)	KNN	2	320 × 240	93.8
	BlueSUAB + 2D CWT [29]	SVM	2	1920 × 1080	93.3
	Visible vegetation index [29]	Thresholding	2	1920 × 1080	58.3
	Green-red vegetation index [29]	Thresholding	2	1920 × 1080	67.6
	Lab + 2D CWT + optical flow [30]	SVM	2	1920 × 1080	96.1
	RGB + entropy [31]	SVM	2	1920 × 1080	94.9
	RGB [31]	SVM	2	1920 × 1080	92.7
	HSV [31]	SVM	2	1920 × 1080	87.3

this region. For a fair comparison, the same types of color and texture features in the clustering learning approach are used. (2) The LeNet-5 approach that uses 64 × 64 regions resized using a fixed ratio method (see Table 3.11). (3) The pixel characteristic approach [11], which constructs a set of pixel-level statistic color features to represent vegetation characteristics and performs vegetation segmentation using an SVM classifier. (4) Approaches that use three opponent color channels $O_1O_2O_3$, their first three moments, and three classifiers—ANN, linear SVM, and KNN. It can be seen that the clustering learning approach outperforms all benchmark approaches and achieves the highest accuracy. The class-semantic textons have significantly higher accuracy (i.e. 19.2%) than using generic textons, confirming the benefit of generating a set of textons specific to each class. Surprisingly, the LeNet-5 produces slightly lower accuracy than the clustering learning approach, and this may be partly due to information lost in the image resizing process. The clustering learning approach also achieves 2.9% higher accuracy than the pixel characteristic approach and nearly the same or higher accuracies than the approaches using $O_1O_2O_3$ color channels and their moments with an SVM, ANN or KNN classifier. The results indicate a superior performance of the clustering learning approach compared with state-of-the-art approaches.

Following [29–31], ten-fold random cross-validations are conducted to obtain an average accuracy on the Croatia roadside grass dataset. The clustering learning approach outperforms the three approaches using fusion of five color channels (i.e. BlueSUAB) and 2D CWT texture features, using threshold-based visible vegetation index, and using threshold-based green-red vegetation index respectively in [29], and the approaches using RGB and HSV color features in [31]. Our approach has slightly lower accuracy (i.e. 1.1 and 2.3%) than the approaches using fusion of RGB and entropy [31], and using fusion of Lab and 2D CWT features in regions of interest detected using optical flow [30], but its performance is obtained using images with a much lower image resolution (320 × 240 vs. 1920 × 1080 pixels). Thus, the clustering learning approach is able to produce performance comparable to state-of-the-art approaches using low-resolution data, which is critical for real-time processing in real-world applications.

3.3.4 Summary

This section has presented a class-semantic texton based clustering learning approach for vegetation segmentation in natural roadside images. It learns class-semantic color-texture textons for an effective representation of class specific features from the training data, and then projects features of all pixels into the learnt textons. A superpixel based collective classification strategy is used to label each superpixel by aggregating the combined occurrence of color-texture textons. Experiments are conducted to investigate the optimal values of several key parameters of the clustering learning approach, which achieves the highest accuracies of 78.9 and 74.5% on two real-world datasets and promising results on a set of real-world videos and the public Croatia roadside grass dataset. The results indicate that shadows of objects and shining pose the biggest challenges for robust segmentation of vegetation, leading to overlap in the color between shadow and tree stem, as well as between brown grass and soil pixels. Furthermore, tree pixels are also prone to be misclassified as road. For accurate segmentation in natural conditions, it is desirable to consider features robust against lighting variations for the segmentation of these objects. The clustering learning approach still can be extended by incorporating statistical features over regions [32] to generate more robust descriptors of objects, as only pixel-level color and texture features are currently considered. In addition, global and local contextual features (e.g. object co-occurrence statistics) can be added to further improve the performance of the approach.

3.4 Fuzzy C-Means Learning

3.4.1 Introduction

Although clustering learning algorithms have shown promising results on object segmentation in roadside data, one drawback of them is that they directly classify the input data into distinct clusters of object categories without considering the fact that the data can potentially belong to multiple clusters. To handle this issue, fuzzy clustering learning algorithms introduce a membership function which allows a sample being assigned (or classified) to different clusters simultaneously, and also provide an indication of the degree of the sample belonging to each cluster. Fuzzy clustering learning algorithms are, therefore, able to more realistically reflect real-world situations and provide more robust and meaningful results for image classification problems even under noisy conditions. As one of the most widely used fuzzy clustering algorithms, the Fuzzy C-Means (FCM) algorithm is an extension to the traditional C-means clustering and has demonstrated good performance in computer vision tasks. It is expected to be suitable for robustly handling various environmental effects for object segmentation on the natural data [33].

This section presents a fuzzy clustering and wavelets based learning approach [34] for roadside object segmentation. Wavelets are used for pre-processing the data to remove blobs like objects having different intensities. The image is then subjected to fuzzy clustering to separate regions of interest and finally extract objects based on color and shape features. Wavelets have also been used previously for object classification [35, 36], but they are designed for good quality images and primarily for recognizing road signs.

3.4.2 *Fuzzy C-Means Learning Approach*

Figure 3.17 illustrates the framework of the wavelet based FCM learning approach [34] for road object segmentation, such as road, sky, and road signs. At the initial stage, the image is subjected to pre-processing (e.g. de-composition and de-noising) by wavelets. The potential regions of interest are then extracted based on the FCM algorithm and fed to an ensemble of neural network classifiers for identifying road objects.

3.4.2.1 Pre-processing

The main task is to pre-process the input image, pass it to FCM clustering, and obtain relevant regions of road objects. Because the image can be of different sizes and resolutions, it is firstly resized to a fixed size of 250 $\times$ 250 pixels, and converted from an original RGB space into an HSI space. The image is further subjected to wavelet pre-processing and fed to fuzzy clustering.

We firstly briefly introduce the wavelets, which are actually a function that divides a signal into multiple individual components. A frequency value is assigned to each component. The frequency can then be studied using the resolution property of wavelets. Wavelet transforms are far better than Fourier transform functions, which can have discontinuities and are unpredictable along the time axis with respect to frequency. In the wavelet based FCM learning approach, CWTs are used as they are efficient in detecting local features of 2D objects. CWTs contain small wavelets with small time durations. Computation of wavelet transforms in a signal

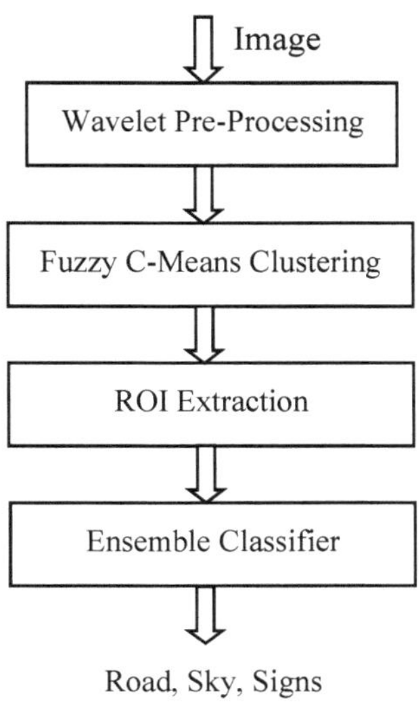

Fig. 3.17 Framework of the wavelet based FCM learning approach

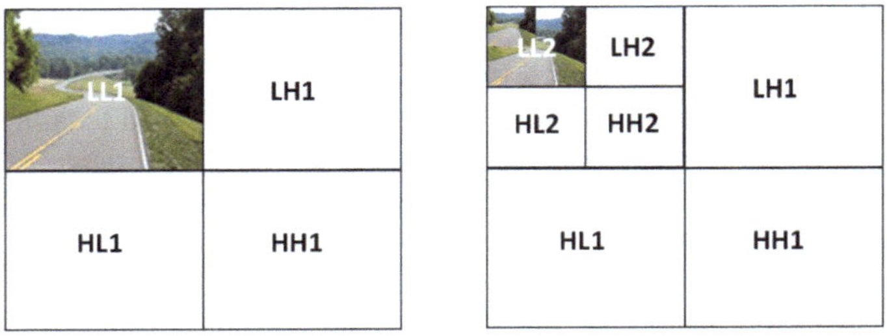

(a) First level decomposition (b) Second level decomposition

Fig. 3.18 First and second level decompositions in CWTs

involves filtering recursively as well as sub-sampling. The input image is decomposed at a certain level such as the first or second level as shown in Fig. 3.18.

The use of wavelets is to improve the quality of images by removing noise based on their multi-resolution property. Wavelet denoising of an image aims to recover the image distorted by noise based on either linear or non-linear filtering. Linear filtering uses low pass filters to cut off high-frequency noise, while non-linear methods use statistic filters. Wavelets use the principle that noise mostly belongs to a high frequency, which can be removed by suppressing the high frequency. This processing segments out blobs like objects of different intensities and shapes.

3.4.2.2 FCM Clustering

Clustering aims to learn the structure of an unlabelled data in an unsupervised manner. The FCM clustering is one such classical method which groups similar training data samples into clusters and learns the fuzzy memberships of each test sample to these clusters. This clustering is done by recursively reducing a cost function, and the fuzzy membership of an instance denotes the degree by which an instance represents different classes. The advantage of FCM is that it can retain more information than other clustering methods.

Unlike the traditional clustering algorithms which create data clusters based on the similarity of data patterns and place similar patterns in the same cluster, fuzzy clustering creates partitions where the data is allowed to belong to different clusters. The FCM algorithm [33] is based on the minimization of the following function:

$$J_m(U, Y) = \sum_{k=1}^{n} \sum_{j=1}^{c} (u_{jk})^m E_j(x_k) \tag{3.26}$$

where, $U = \{x_k | k \in [1, n]\}$ represents the training set with n unlabelled samples, $Y = \{y_j | j \in [1, c]\}$ represents a set of cluster centres, $E_j(x_k)$ is the dissimilarity measure of the sample x_k and the centre y_j of a specific cluster j, u_{jk} represents the fuzzy partition matrix, and $m \in (1, \alpha)$ is a fuzzy parameter.

The value of J_m with respect to Y can be minimized using:

$$\sum_{j=1}^{c} (u_{jk}) = 1 \tag{3.27}$$

It is ensured that pixels closer to the cluster centroids have high values, while those far from the centroids have low values. The membership value in FCM is based on the distance of pixels to cluster centres.

3.4.2.3 Region of Interest (ROI) Extraction

Once the image was processed by the FCM, it is then subjected to a pixel based search process, which uses color features to find the ROI. For road, the ROI is searched at the bottom part of images and the extracted ROI is then classified as road or non-road. In a similar way, the search process for sky is performed at the top part of images, and the extracted ROI is classified as sky or non-sky. Color features are then obtained from the extracted ROI. Traffic signs are obtained by a template matching process, which accounts for the intensity of pixel concentration. Color features related to traffic signs are used for assisting the matching process. The segmented areas are matched with attributes of road signs like color and size. Finally, we obtain ROIs or candidate road objects.

3.4.2.4 Object Classification

An ensemble of neural networks is used as the final classifier for distinguishing the extracted ROIs. Ensemble classifiers have the advantage of combining decisions from multiple classifiers to reach more robust and accurate results. Neural networks during training act differently due to different initial parameters like weights, hidden neurons, etc. on the same dataset. As a result, each network can result in different classification errors, and combining them can reduce the error and improve the accuracy. We used the MLP to form the ensemble classifier and each MLP is trained separately using color features extracted from ROIs. Parameters of MLP classifiers are changed for obtaining a diversity. At the classification stage, each MLP classifies the obtained ROIs into two classes denoted by $Y_k, k = 1, 2$ so that a class Y_1 corresponds to sky and class Y_2 corresponds to non-sky for instance. Five MLPs are developed to classify the obtained ROIs into different object categories. The final result is obtained by combining the results of all MLPs using a majority voting strategy.

3.4.3 Experimental Results

3.4.3.1 Evaluation Metrics and Systematic Parameters

The evaluation dataset is the natural road image dataset that was collected using a frontal view camera. We estimate the number of objects correctly classified using the FCM learning approach. Two measurements are used: (1) Correct Recognition Rate (CRR), which indicates the number of all correctly classified objects divided by the total number of objects, and (2) False Recognition Rate (FRR), which indicates the percentage of objects wrongly classified to the total number of objects.

Systematic parameters: three-layered MLP classifiers with one hidden layer are used and trained using the backpropagation algorithm. The parameters used for training are as follows: (1) learning rate = 0.01; (2) momentum = 0.2; (3) no. of iterations = 60; and (4) RMS value = 0.001. The best parameter settings for different datasets are selected in a trial and error process.

Fig. 3.19 A sample set of pre-processed images in the natural road image dataset

3.4.3.2 Experimental Results

Figure 3.19 shows the results of an example set of pre-processed images. The final set of road objects extracted using the wavelet based FCM approach is shown in Fig. 3.20, including signs, sky and road.

Good classification is indicated by a high CRR and a low FRR. We have compared the wavelet based FCM approach with the FCM approach [37]. The results obtained are described in Table 3.13, showing the advantage of incorporating wavelets for data pre-processing in increasing the CRR and reducing the FRR. The FCM method performs well on images free of noise, but is sensitive to noise. The wavelet based FCM approach uses wavelets to remove noise and the accuracies obtained are higher than those of the FCM approach.

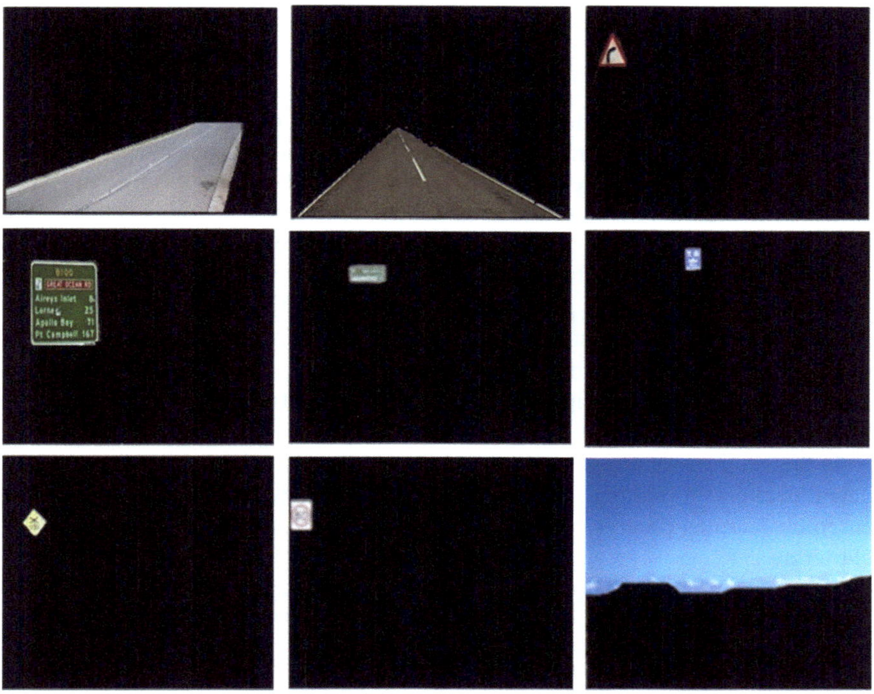

Fig. 3.20 A sample set of extracted road objects

Table 3.13 Performance (%) comparisons of the wavelet based FCM approach with the FCM approach

	FCM approach		Wavelet based FCM approach	
	Road	Sky	Road	Sky
CRR	88.4	98.1	94.5	98.8
FRR	2.1	2.8	0.01	0.02

3.4.4 *Summary*

This section has presented a wavelet based FCM learning approach for object segmentation from roadside images. Identification of ROIs in road images is completed using a wavelet based FCM clustering technique, and classification of road objects is performed using an ensemble of MLP classifiers. Experiments conducted on real-world roadside images show that recognition accuracies of 94.5 and 98.8% were obtained for road and sky respectively. Experimental comparisons with an existing FCM approach indicate big improvements in classification accuracy using the wavelet based FCM approach.

3.5 Ensemble Learning

3.5.1 *Introduction*

Ensemble learning is a method that combines multiple models of the same type or different types to obtain a final output by adopting a fusion strategy such as majority voting and plurality. Different sets of parameters can be given to these models to achieve a diversity of the results. The advantage of ensemble learning is that it takes into consideration decisions from multiple individual models so that both classification accuracy and robustness of the ensemble learning system can be greatly improved in most cases. In general, each model can produce accurate results from only a proportion of the test data, but a combination of multiple models is capable of achieving higher performance by considering all decisions from individual member models. It is anticipated that an ensemble learning approach can achieve better results in roadside video data analysis.

In this section, we present an ensemble learning approach [38] based on multiple neural networks to segment and classify roadside images into different objects. The ensemble learning process combines the decisions of multiple classifiers created by clustering.

3.5.2 *Ensemble Learning Approach*

Figure 3.21 shows the framework of the approach [38] for generating an ensemble of classifiers based on the concept of clustering and fusion. The first task is to cluster the input images into multiple segments and employ a set of base classifiers to learn the decision boundaries among patterns in each cluster. This clustering process partitions a dataset into segments that contain highly correlated data points, which tend to be closer to each other geometrically. When patterns from multiple classes overlap within a cluster, these data points are difficult to be classified.

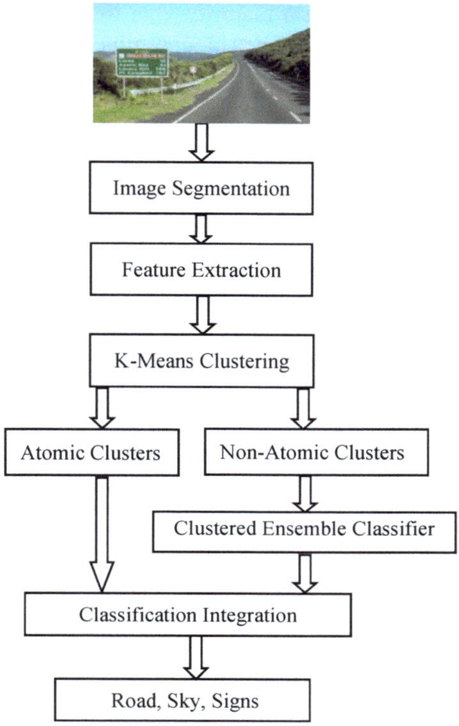

Fig. 3.21 Framework of the ensemble learning approach

Applying clustering onto datasets associated with a class will generate two types of segments: atomic and non-atomic. An atomic cluster is with patterns that belong to the same class, whereas a non-atomic cluster has patterns from multiple classes.

After the clustering process, classifiers are trained based on the patterns of non-atomic clusters and the class labels are assigned for atomic clusters. The class label of a test pattern is predicted by finding a suitable cluster based on its distance from the cluster centre, using the corresponding class label for atomic clusters, and then using a suitable classifier for non-atomic clusters. Clustering helps in identifying those patterns that are difficult to be classified. Once the clustering operation was performed and clusters were identified, a neural network classifier is trained for each cluster.

In K-means clustering, the labelling of patterns to a cluster can be different based on the initial states of cluster centres, where the number of K-means clusters may be different to the actual number of clusters in the data. If multiple clusterings are performed with different seeding points, patterns might be assigned to a different cluster each time. When a new clustering operation is performed with different initial seeds, it is called layering and these clusters form a layer. This cluster labelling is different from one layer to the next. A classifier can be trained on those non-atomic clusters for each layer and the results of all classifiers can be combined together using a majority vote to form an ensemble. The layers provide a way of keeping a diversity in ensembles and making it easier to classify non-atomic patterns.

3.5.2.1 Image Segmentation

For image segmentation, we use color feature based clustering which takes into account characteristic features related to changes in color components. The first step is to measure color features. Firstly, road images are segmented into two color channels: white and non-white using K-means clustering with $K = 2$. The segmentation produces white segments for lanes, sky, dry vegetation and road signs, and non-white segments for road, colored road signs and green vegetation. The potential road objects are located by their positions in the image. Road extraction is done by block based feature extraction at the bottom part of the image, while sky regions are separated by confining the search to the top part of the image. The sky is extracted from the white segment by searching blobs which are connected to the top edge of the image.

3.5.2.2 Feature Extraction

The segmented image is then used for block based feature extraction. A block size is firstly defined which divides the image into a number of equal blocks. The ensemble learning approach uses a block size of 64×64 pixels.

For road segments, the image is divided into blocks only at the bottom part of the image and each block is labelled into three classes including road, non-road, and background. For sky segments, the image is divided into blocks at the top of the image, and each block is labelled as sky, non-sky, or vegetation. Feature extraction is then performed in each block. Four types of road signs are extracted from the image using color ranges, including green signs, light blue signs, yellow signs, and speed signs. Once the boundaries of regions are extracted, they are further filtered by comparing the signature of each blob with those obtained from the reference shapes.

3.5.2.3 Clustering and Ensemble Classifiers

All road images are clustered using the K-means clustering algorithm, which produces both atomic clusters where only one class membership exists and non-atomic clusters where more than one classes present. A neural network is then trained on non-atomic clusters which produces a layer. This process is repeated for a number of clustering operations with different initial cluster seeding points. Each classifier layer is trained to recognize decision boundaries for non-atomic clusters.

After the training operation is completed, the network is applied to the test data. During testing, the ensemble classifier evaluates as to which class a test pattern belongs in two steps. In the first step, the cluster membership is determined based on the pattern's distance to the cluster centroid. If the pattern belongs to an atomic cluster, then the class label for that cluster is returned. If the pattern belongs to a non-atomic cluster, then the class label from the trained network is obtained. Finally, a majority vote is used to fuse decisions from ensemble classifiers.

Figure 3.22 graphically explains the way of creating layers that occur by varying the seeds of cluster centers. In Fig. 3.22a, three clusters are defined and the class membership indicates that two atomic clusters and one non-atomic cluster have been formed. The non-atomic cluster contains multiple classes so a neural network classifier needs to be trained on this cluster. The atomic clusters are easier to be classified and their class labels signify the test patterns. In Fig. 3.22b, the cluster seeding is changed and the layer produces three clusters. In this case, cluster memberships are different substantially to the previous layer. This difference in patterns creates a diversity during neural network training, which improves the performance of the ensemble classifier.

The use of neural classifiers on different clusters provides a method by which a diversity in the classification outputs of neural networks is achieved. As the neural network is used in different cluster groupings, a clustered ensemble is formed by combining all outputs.

3.5.2.4 Classification Integration

The outputs obtained from different neural networks trained on each clustered layer are integrated by a majority vote, which is anticipated to be able to improve the overall output compared to using each individual classifier. The majority voting chooses a class label that has the highest number of votes across all classes.

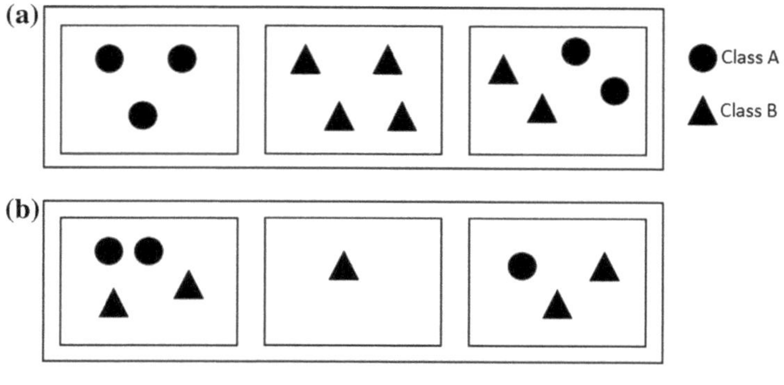

Fig. 3.22 Layering on the data which are clustered into **a** two atomic clusters and one non-atomic cluster, and **b** three clusters

3.5.3 Experimental Results

3.5.3.1 Experimental Settings

The evaluation dataset is the natural road image dataset. Images in this dataset are representative of segmentation problems that arise in real-world road conditions. There are no established criteria for measuring the segmentation performance. Although the work [31] described various methods to measure segmentation performance, but it did not specify any standard criteria. Four measures are used for evaluating the ensemble learning approach: (1) correct recognition rate, which is the number of all correctly classified objects divided by the total number of all objects; (2) number of lost objects, which relates to the number of objects that are not classified correctly; (3) number of maximum scores, which indicates the times that a method achieves a maximum score; and (4) false classification rate, which represents the number of objects wrongly classified divided by the total number of objects classified.

3.5.3.2 Benchmark Approaches

This section describes three benchmark approaches that are used in the experiments for road object classification. They serve as benchmark approaches for performance comparisons with the ensemble approach.

(1) SVM approach [39], which extracts road objects using an SVM classifier that determines a linear hyperplane with the largest margin in a high-dimensional feature space. We extract a feature vector at each image pixel which is classified by the trained SVM classifier.
(2) Hierarchical segment learning [32], which uses hierarchical segment extraction and neural network based classification of segmented objects. Objects such as sky, road, sign and vegetation are extracted at the hierarchical stage and classified using a neural network classifier. A total of $(960 \times 1280)/(64 \times 64)$ segments are used in each image with a block size of 64×64 pixels. The images are then subjected to clustering and feature extraction.
(3) Clustering based neural network [4], which combines clustering and a neural network classifier for the segmentation of road images into different objects. It generates clusters for each class and uses those clusters to form subclasses for each extracted segment. An integration of clusters in the classification is employed to improve classification accuracy of the system. The classifier is MLP with a single hidden layer and is trained on each cluster and the results are integrated. The classifier is trained using the backpropagation algorithm with the following parameter settings: learning rate = 0.01; momentum = 0.2; no. of iterations = 55; and RMS goal = 0.01. The best parameter settings on datasets are found by trials with respect to the error.

3.5.3.3 Performance Results

Using the above measures, the performance of the ensemble learning approach is compared with those of three benchmark approaches as shown in Table 3.14. The highest classification rate is obtained using the ensemble learning approach, which is 91.2% accuracy. There are more than 2.8% accuracy improvements compared to benchmark approaches. The number of objects lost and falsely detected using the ensemble approach is also lower than benchmark approaches. Figure 3.23 shows a sample set of extracted road objects using the ensemble approach, and Fig. 3.24 compares the performance between all approaches based on different measures, showing a better overall performance of the ensemble learning approach.

Table 3.14 Performance comparisons of the ensemble learning approach with three benchmark approaches

Measure	SVM	Hierarchical	Clustering	Ensemble
Correct rate (%)	80.2	81.5	88.4	91.2
Lost	4	5	3	2
Max	6	9	4	12
False rate (%)	2.34	3.4	2.1	0.00

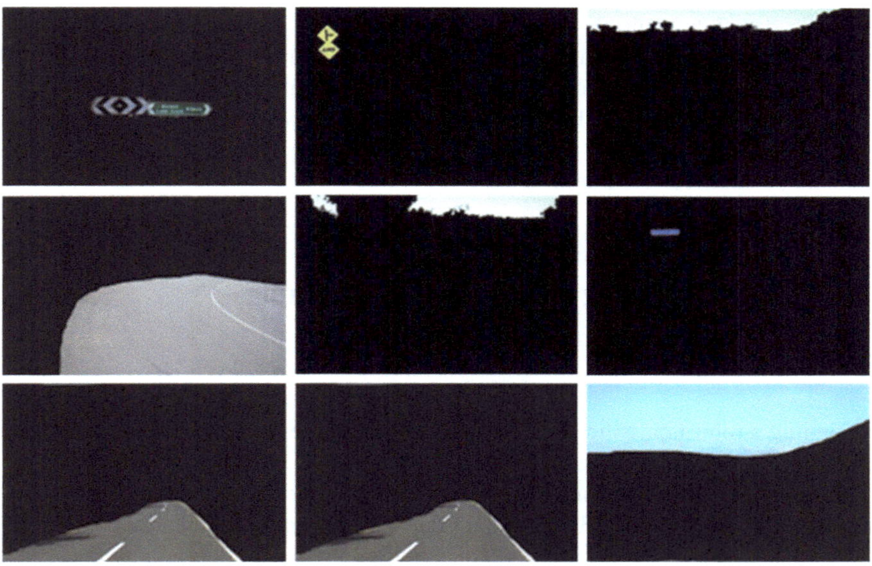

Fig. 3.23 A sample set of extracted road objects using the ensemble learning approach

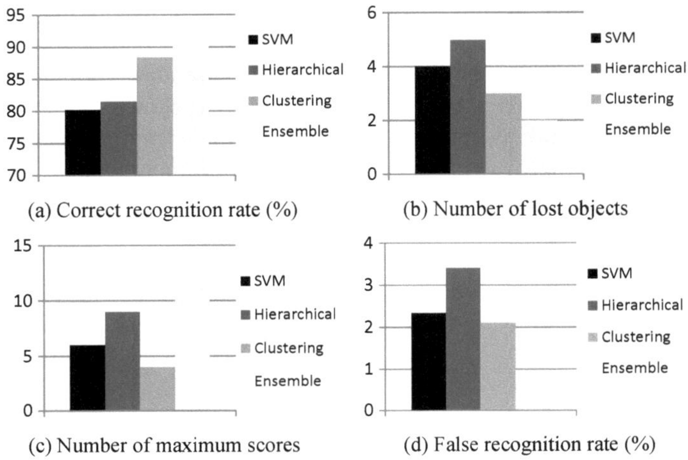

Fig. 3.24 Comparative analysis of the ensemble learning approach with three benchmark approaches

3.5.4 Summary

In this section, we have presented a neural network based ensemble learning approach for road object detection. The approach uses different seeding points to partition data into multiple layers and then generates clusters indicating different patterns at each layer. A neural network classifier is trained on each cluster in each layer, and the outputs of an ensemble of clustered networks are fused using a majority vote. Experimental results confirm big improvements in the classification rate over benchmark SVM, hierarchical and clustering approaches for road object detection with a correct rate of 91.2%.

3.6 Majority Voting Based Hybrid Learning

3.6.1 Introduction

Designing a robust classification algorithm is one of the most crucial components in building an automatic system for video data analysis [40]. There are many algorithms proposed for solving the classification problem including KNN [41], AdaBoost [42], ANN [43], SVM [44], Wavelet based techniques [45], etc. However, none of these algorithms can guarantee any form of optimality in terms of proper classification accuracy in realistic environments. In recent years, the hybrid approach of combining multiple classifiers [46, 47] rather than a traditional way of using a single classifier has received increasing attention due to its effectiveness and robustness in classification problems, particularly in real-world scenarios [48]. However, the majority of existing approaches to vegetation segmentation focus on using a single classifier, and few studies have explored roadside vegetation segmentation using video data captured by vehicle-mounted cameras.

This section describes a majority voting based hybrid approach [49] to combine multiple classifiers for dense versus sparse vegetation classification. Three types of classifiers including ANN, SVM and KNN are specifically selected to improve the diversity, which helps achieve better performance. The strengths of the majority voting based approach include the incorporation of multiple classifiers with majority voting and a new feature extraction technique. However, because the approach combines multiple classifiers, it requires a longer time to process all classifiers and thus poses a challenge for real-time processing.

3.6.2 Majority Voting Approach

The majority voting based hybrid approach for discriminating dense versus sparse grasses is designed based on the assumption that vegetation types can be learnt and distinguished by a fusion of multiple texture features with multiple classifiers. Figure 3.25 shows an overview of the approach, which consists of five stages including data acquisition, image pre-processing, training of base classifiers, fusion of classifier decisions, and accuracy calculation.

3.6.2.1 Image Pre-processing

The pre-processing step is designed to prepare an input image so that it can be directly used for feature extraction. This step includes filtering, color space conversion, and resizing.

(1) Median filtering. To remove noise from the input image, median filtering is applied to the image, which generates a smoothed and clean version of the image.
(2) RGB to gray conversion. To support feature extraction in gray-scale images, all color images are converted to gray scales by taking an average value of R, G, and B as the intensity value of each pixel.
(3) Image resizing. The images have a resolution of 900 $\times$ 500 pixels during image acquisition. To reduce the computational time, all images are resized to a resolution of 200 $\times$ 200 pixels.

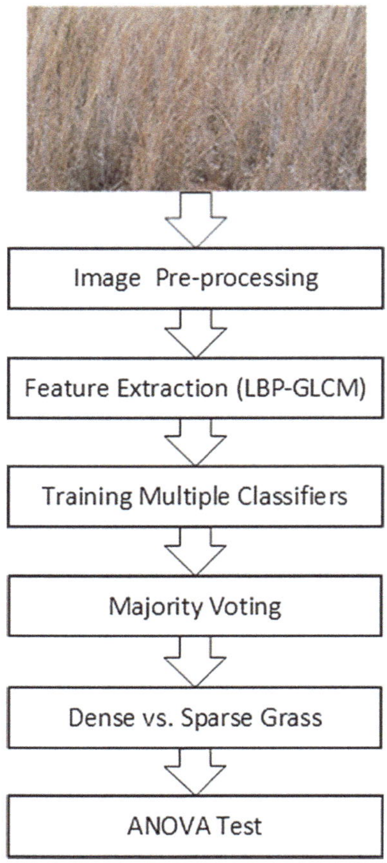

Fig. 3.25 Framework of the majority voting based hybrid approach

3.6.2.2 Feature Extraction

Feature extraction is one of the most crucial steps for vegetation segmentation. It is observed that dense and sparse grasses can be visually separated based on their smoothness levels in the texture and differences in grass depths. Based on this observation, the majority voting approach presents a texture extraction technique to obtain co-occurrence of binary patterns based on a combination of LBP and GLCM. To be specific, the LBP operator is firstly applied on gray scale images, which is followed by applying GLCM to generate a texture feature vector.

The LBP is a gray-scale and rotation invariant texture feature extractor, which extracts a histogram of integer labels from the image. The LBP operator forms the labels of image pixels by thresholding a neighborhood of each pixel (e.g. 3×3 pixels) with the center value. For each pixel, a binary value is obtained by concatenating all these results binomially in a clockwise direction, and it is then assigned to the center pixel. The LBP code of a pixel (x_c, y_c) is obtained by:

$$LBP_{P,R}(x_c, y_c) = \sum_{P=0}^{P-1} s(i_p - i_c)2^p \quad (3.28)$$

$$s(x) = \begin{cases} 1, & x \geq 0 \\ 0, & x < 0 \end{cases} \quad (3.29)$$

where, i_c represents the gray value of the centre pixel (x_c, y_c), i_p is gray values of its neighbours, P is the number of neighbors and R is the radius of the neighborhood. For a neighbor that does not fall exactly on a pixel position, its value is estimated using a bilinear interpolation.

After calculating the LBP code for each pixel (x, y) in an image of $M \times N$ pixels, we get an encoded image representation. A histogram H is obtained from the encoded image by counting the occurrences of the LBP codes using:

$$H(b) = \sum_{x=1}^{M} \sum_{y=1}^{N} f(LBP_{P,R}(x, y), b), f(a, b) = \begin{cases} 1, & a = b \\ 0, & a \neq b \end{cases} \quad (3.30)$$

where, b is the LBP code value. The resulting histogram H is used as the feature vector that describes texture of the image, which is further used as inputs to the GLCM algorithm.

In the next phase, we use the GLCM algorithm to extract texture features. The GLCM is a gray-scale texture primitive that describes the spatial structure of the local texture in an image. Gray-level co-occurrence matrices indicate the frequency that a pixel with a gray intensity value is horizontally adjacent to another pixel at a given offset. A matrix for each gray-scale version of the image is constructed.

3.6.2.3 Training Base Classifiers

For classifying dense versus sparse grass images, three classifiers namely SVM, ANN and KNN are used.

(1) SVM classifier. The first classifier is SVM, which is designed to find the best separation between classes. Let $S = \{(x_i, y_i) | x_i \in R^n\}$ and $y_i \in \{1, 2\}$ indicate two class labels for training. A class "1" represents dense grasses, while "2" represents sparse grasses. Three kernel functions including linear, polynomial and RBF are considered.
(2) ANN classifier. The second classifier is a three-layer feed-forward ANN. Let assume $u = [u_1, u_2, u_3, \ldots u_p]^T$ constitute the input feature vector, and $y = [y_1, y_2, y_3, \ldots y_m]^T$ be the output vector, where p represents the number of elements in u, i.e. $p = 110$, and m denotes the number of classes, i.e. $m = 2$. The ANN is trained with different numbers of hidden units (i.e. 6, 10, 12, 15 and 20) and

iterations (i.e. 500, 1000 and 3500) repeatedly until the RMS error on the training samples becomes below a preset value using the backpropagation algorithm.

(3) KNN classifier. The third classifier is KNN, where a test object is classified to the class label of the closest training example in a feature space. KNN has two parameters to be tuned: *K* and the distance metric. *K* is usually set to an odd number to avoid draw votes. Three values of K are tested including 5, 7 and 9, and two distance metrics are used including the Euclidean and City Block. One major problem of KNN is that the class with more frequent training samples would dominate the prediction results of test samples. To overcome this problem, our experiment adopts the same number of images for two classes.

3.6.2.4 Majority Voting Over Classifiers

After training all three classifiers using the same image descriptor, the decisions from them are combined to reach a final classification decision using a majority vote. The class which has the majority of votes wins, where at least two classifiers provide the same result. Finally, a test image is labelled as dense or sparse.

3.6.3 Experimental Results

This section provides experimental results on the cropped grass dataset. Two experiments are carried out. The first phase chooses the best parameters for each classifier and the second phase applies 5-fold cross validations to obtain the classification rate with the chosen parameters. A total of 110 images consisting of 60 dense and 50 sparse grasses are used. All images are randomly partitioned into five equal subsets with an equal number of instances for dense or sparse grasses. In each validation, one subset is used for testing and the remaining subsets for training. The above process is repeated for five times to produce an average classification rate.

Table 3.15 shows the results obtained using the SVM classifier, including training and test accuracies for three kernel functions. The linear function achieves the highest accuracies on both training and test data, which are 90% for training and 85% for test, whereas the polynomial and RBF functions show lower accuracies, i.e. the training and test accuracies achieved are 85 and 80% respectively using the polynomial and 80 and 80% respectively using the RBF.

Table 3.15 Classification accuracy (%) using the SVM classifier

Kernel function	Linear	Polynomial	RBF
Train Acc.	90	85	80
Test Acc.	85	80	80

Table 3.16 lists the classification results of SVM using 5-fold cross validations with a linear kernel function.

Table 3.17 lists the results obtained versus different parameters in the ANN classifier. The ANN produces the highest accuracy with the following parameters: number of hidden units = 12, number of iterations = 3500, learning rate = 0.01, momentum = 0.15 and RMSE = 0.0001. The training and test accuracies are 90 and 85% respectively. This implies that ANN is able to achieve performance similar to SVM on the condition that proper parameters are chosen for it.

Table 3.18 shows the classification accuracy of ANN using 5-fold cross validations with the chosen parameters.

Table 3.19 presents the classification results obtained using KNN. The training and test accuracies using different K values are compared to select the best value. In this case, 90 and 20 images are used for training and test respectively. The highest accuracies on the training and test datasets are obtained using $K = 7$. The accuracies

Table 3.16 5-fold cross validation results (%) using a linear SVM classifier

Fold	1	2	3	4	5	Overall
Acc.	95.5	90.9	95.5	86.4	91.0	91.8

Table 3.17 Classification accuracy (%) using the ANN classifier

Exp#	Hidden unit	Iteration	RMSE	Train Acc.	Test Acc.
1	6	500	0.0003	80	75
		1000	0.0004	75	75
		3500	0.0001	80	75
2	10	500	0.0005	80	80
		1000	0.0001	80	80
		3500	0.0003	85	85
3	12	500	0.0001	85	80
		1000	0.0002	85	85
		3500	0.0001	90	85
4	15	500	0.0002	85	80
		1000	0.0003	90	80
		3500	0.0001	90	80
5	20	500	0.0004	85	80
		1000	0.0003	85	80
		3500	0.0002	85	80

Table 3.18 5-fold cross validation results using the ANN classifier

Fold	1	2	3	4	5	Overall
Acc.	90.9	90.9	90.9	95.5	90.9	91.8

Table 3.19 Classification accuracy (%) using the KNN classifier

K Value	5	7	9
Train Acc.	75	85	75
Test Acc.	70	80	70

Table 3.20 5-fold cross validation results using KNN with K = 7

Fold	1	2	3	4	5	Overall
Acc.	90.9	86.4	86.4	90.9	95.5	90.0

for the training and test datasets are 85 and 80% respectively. Although the accuracies obtained using KNN are lower than those of ANN and SVM, the acceptance rate of KNN is close to those of ANN and SVM.

After obtaining the best value of K, we evaluate the KNN classifier in 5-fold cross validations as shown in Table 3.20.

Tables 3.21 and 3.22 summarize the results using the majority voting approach. The results indicate that the approach achieves the highest classification accuracy when a linear kernel function is used for SVM, the number of hidden neurons and iterations are set to 12 and 3500 respectively for ANN and the K equals to 7 for KNN. The highest training and test accuracies are 95 and 90% respectively.

Table 3.23 compares the correct classification rate of the majority voting approach with three individual classifiers. It is observed that fusion of classifiers shows better performance than using individual classifiers, and the highest accuracy is 92.7%. The overall accuracy achieved by the majority voting approach shows its

Table 3.21 Classification accuracy (%) using the majority voting approach

Exp#	SVM	ANN	KNN	Train Acc.	Test Acc.
1	Linear	HN = 12, iterations = 3500	7	95	90
	polynomial	HN = 10, iterations = 3500	5	85	80
	RBF	HN = 15, iterations = 1000	9	80	75
2	Linear	HN = 10, iterations = 3500	9	85	80
	polynomial	HN = 15, iterations = 1000	5	80	75
	RBF	HN = 12, iterations = 3500	7	85	80
3	Linear	HN = 15, iterations = 1000	5	80	75
	polynomial	HN = 12, iterations = 3500	9	80	80
	RBF	HN = 10, iterations = 3500	7	80	75

Table 3.22 5-fold cross validation results using the majority voting approach

Fold	1	2	3	4	5	Overall
Acc.	90.9	95.5	90.9	90.9	95.5	92.7

Table 3.23 Comparisons of classification rates between classifiers using 5-fold cross validations

Classifier	#Total samples	#Misclassified samples	Rate (%)
SVM	110	9	91.8
NN	110	9	91.8
KNN	110	11	90.0
Majority voting	110	8	92.7

Fig. 3.26 Examples misclassified as dense (*left*) and sparse (*right*) grasses

best performance overall but some misclassification also occur as shown in Fig. 3.26, where the left sample with sparse grasses is misclassified as dense grasses. A similar misclassification result happens for the right sample, where moderate grasses are misclassified as sparse.

The above results show that SVM and ANN perform similarly with accuracies around 90 and 85% on the training and test data respectively. The results using KNN show 85 and 80% accuracies on the training and test data respectively. Finally, the highest classification accuracies are obtained by using the majority voting approach, which are 95 and 90% on the training and test data respectively. Hence, we can conclude that the majority voting approach that fuses multiple classifiers is more robust for roadside vegetation classification than using individual classifiers.

A single factor ANalysis Of VAriance (ANOVA) test is performed on 5-fold cross validation results which compares the majority voting approach with SVM, ANN and KNN to test if the improvement in classification accuracy of the majority voting approach is statistically significant. The null hypothesis is that there is no statistical variance in classification accuracy between the majority voting approach and individual classifiers (H_0), while the alternative hypothesis is that there is a significant difference (H_1). The value of p indicates whether one should reject the null hypothesis in favor of the alternative.

Tables 3.24 and 3.25 present the results of the ANOVA test. Observed from Table 3.24, the majority voting approach has the highest accuracy. It also has much

Table 3.24 Summary of the single factor ANOVA test

Group	Count	Sum	Average	Variance
SVM	5	459.11	91.822	14.44217
NN	5	459.05	91.81	4.1405
KNN	5	449.97	89.994	14.45538
Majority voting	5	463.6	92.72	6.21075

Table 3.25 Results of the ANOVA test

Variation	SS	Df	MS	F	P-value	F-Crit
Between	19.6314	3	6.543818	0.6669	0.584529	3.2388
Within	156.9952	16	9.8122			
Total	176.6267	19				

less variance in the accuracy than SVM and KNN, but slightly higher variance than ANN. As the p value is larger than 0.1 in Table 3.25, we can conclude that there is no significance batch effect at the 0.1 level of significance, and the null hypothesis cannot be rejected. Although the majority voting approach achieves the highest classification accuracy, the differences in accuracy between it and benchmark approaches are not statistically significant.

3.6.4 Summary

In this section, we have presented a majority voting based hybrid approach for classifying dense and sparse grasses in natural roadside images. The original images are firstly passed through several pre-processing steps to make them ready for feature extraction. Considering the fact that, pixel intensity values in dense grass regions often remain closely compared to big differences in sparse grass regions, LBP based binarization is executed to generate histogram based texture features, which are further processed using the GLCM algorithm to extract the texture features for each image. Finally, fusion of three classifiers (i.e. ANN, SVM, and KNN) using a majority voting is applied to classify dense and sparse grasses.

Experiments are conducted on the cropped grass dataset using 5-fold cross validations, and the results show that the majority voting based hybrid approach outperforms all individual classifiers with 92% accuracy. However, an ANOVA test is performed which confirms that the performance of the majority voting approach is not statistically significantly better than those obtained using individual classifiers. Although the experimental results are quite encouraging, the approach still can be further improved. It has not considered the whole roadside image and thus further investigation is still needed to segment grass regions from images. All test data was captured in good lighting conditions, and the approach may fail to handle environmental effects such as shadows and bad weather conditions. Only dense and sparse grasses are considered, and it is necessary to conduct tests on a large dataset of grasses and mixed vegetation like trees and shrubs.

3.7 Region Merging Learning

3.7.1 Introduction

Object segmentation can be treated as a region merging problem, where small regions with similar characteristics are progressively merged into larger regions, and each merged region can then be labelled into a class category. Most existing region merging methods [5, 25] select a small set of initial seeds with a high level of confidence in class labels and then iteratively merge all pixels (or superpixels) into the most similar neighboring seeds. These methods generally require no training and focus only on local characteristics of the test data, and thus they are suitable for adaptive object classification. However, a major drawback of them is a high dependence on initial seed selection, which often result in low reliability in natural conditions. Existing solutions include manual selection of initial superpixel seeds [50] and selecting initial seed pixels based on Gaussian probability density functions [25], but they suffer from either human interventions or heavy computation burdens.

This section presents a Spatial Contextual Superpixel Model (SCSM) [51], which combines pixel based supervised class-specific classifiers and superpixel based unsupervised region merging for robust roadside object segmentation. The SCSM includes an adaptive superpixel merging algorithm to overcome the dependence on initial superpixel seeds by accounting for both general characteristics of objects in the training data and local characteristics of a test image, such as lighting conditions and vegetation types (an example shown in Fig. 3.27). Thus, the SCSM is capable of being automatically adapted to the local content of the test image, and thus is expected to produce more robust classification.

3.7.2 Region Merging Approach

3.7.2.1 Framework of Approach

Figure 3.28 depicts the framework of the SCSM approach [51], which takes a roadside image as an input and assigns each part of the image into an object category. There are two major processing steps:

Fig. 3.27 Image examples illustrating the necessity of considering local characteristics of objects (i.e. tree leaf) in a test image. The two regions of tree leaves have substantially contrasting differences in pixel intensities, and thus a classification system trained using features extracted from the left region may not work well in the right region

(1) The image is segmented into a set of superpixels, from which Pixel and Patch Selective (PPS) features are extracted for training multiple class-specific ANN classifiers. The PPS features are designed specifically to reduce noise presented in boundaries of objects. The probabilities predicted by the ANN classifiers are then aggregated over all pixels within each superpixel to form Contextual Superpixel Probability Maps (CSPMs), with the consideration of several spatial constraint models.
(2) Superpixel classification is performed based on the CSPMs to assign each superpixel to a class category that has the highest average probability over all pixels within the superpixel. To account for both local spatial information between neighboring superpixels and global contextual information in a test image, a superpixel merging method is presented to obtain category labels of superpixels with low probabilities. The method compares the similarities of those low-probability superpixels with their neighbors and with a set of seed superpixels of high probabilities. This process helps to enforce more consistent class labels in a local spatial neighborhood and achieve higher classification accuracy in the whole image.

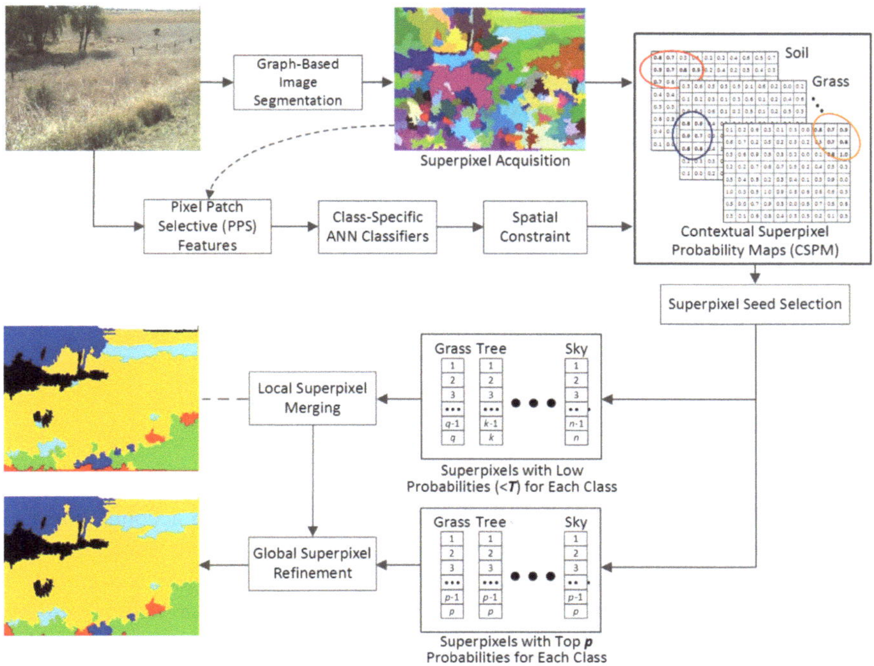

Fig. 3.28 Framework of the spatial contextual superpixel model (SCSM)

3.7.2.2 Superpixel Generation

The first task of the SCSM is to segment the input image into a group of local superpixels. This process converts the classification problem from tens of thousands of single pixels to hundreds of large homogeneous regions, which can significantly reduce the complexity of the object categorization process to a more manageable level. The segmented superpixels form the foundation for PPS feature extraction and superpixel seed selection. The popular graph-based algorithm [23] is employed to segment images into a set of superpixels, and each superpixel is expected to belong to only one class. The segmented superpixels are also expected to consider the contrastive visual differences between objects and preserve most of object boundary information.

3.7.2.3 Feature Extraction

Feature extraction attempts to extract a set of discriminative features from the input data to distinguish different object categories. Two types of features are considered: pixel based features and patch based features, which form the foundation for generating more robust PPS features.

(1) Pixel based features, which are extracted separately from a single image pixel. Color is one of the most widely used pixel based features, but it is still a challenge to select a suitable color space. It is generally recommended that the color space should be approximately uniform to human color perception, as humans are very adept at distinguishing objects. In other words, equal distances in the color space should ideally correspond to equal color differences perceived by humans. Therefore, we select the CIELab color space, which has demonstrated high consistency with human vision perception and high performance of object categorization [52]. In addition, we also include R, G, B color channels to compensate for the information possibly lost in the Lab space. Thus, we obtain a 6-element pixel based feature vector $V^I_{x,y}$ for a pixel $I_{x,y}$ at the coordinates (x, y) in an image I:

$$V^I_{x,y} = [R, G, B, L, a, b] \tag{3.31}$$

(2) Patch based features, which can be extracted by taking into account statistical information in a neighborhood of pixels. Spatial texture information in neighboring pixels plays a crucial role in complicated object recognition in real-world applications. In the SCSM, patch based features are extracted based on color moments [53], which have advantages of encoding both shape and color information of objects, being scaling and rotation invariant, and having high robustness against lighting variations. Because most of color distribution information is represented in low-order moments, the first three moments, including mean, standard deviation and skewness, are used.

Let $I_{x,y}$ be the pixel at the location (x, y) in an input image I, $T_{x,y}$ be the patch centred at $I_{x,y}$ with a height h and a width w, τ be the number of all pixels in $T_{x,y}$, i.e.$\tau = h \times w$, and $I_{i,j}$ be the pixels belonging to $T_{x,y}$, i.e. $I_{i,j} \in T_{x,y}$, the first three moments of $I_{x,y}$ can be expressed by:

$$Mean_{x,y} = \frac{1}{\tau} \sum_{i=x-\frac{w}{2}}^{x+\frac{w}{2}} \sum_{j=y-\frac{h}{2}}^{y+\frac{h}{2}} I_{i,j} \tag{3.32}$$

$$Std_{x,y} = \sqrt{\frac{1}{\tau} \sum_{i=x-\frac{w}{2}}^{x+\frac{w}{2}} \sum_{j=y-\frac{h}{2}}^{y+\frac{h}{2}} (I_{i,j} - Mean_{x,y})^2} \tag{3.33}$$

$$Skw_{x,y} = \sqrt[3]{\frac{1}{\tau} \sum_{i=x-\frac{w}{2}}^{x+\frac{w}{2}} \sum_{j=y-\frac{h}{2}}^{y+\frac{h}{2}} (I_{i,j} - Mean_{x,y})^3} \tag{3.34}$$

The above three moments are extracted for each of L, a, and b channels.

Because the calculation of the color moments does not consider the coordinates of pixels, they have a problem of capturing the spatially structured texture in a local neighborhood. To handle this problem, two additional features are included to indicate the vertical texture orientation, including the differences between means and between standard deviations in two left-right half parts of a patch. In the same way, two additional features are calculated from two top-bottom half parts:

$$Mean_{x,y}^{l,r} = \frac{1}{\tau}\left(\sum_{i=x-\frac{w}{2}}^{x}\sum_{j=y-\frac{h}{2}}^{y+\frac{h}{2}} I_{i,j} - \sum_{i=x}^{x+\frac{w}{2}}\sum_{j=y-\frac{h}{2}}^{y+\frac{h}{2}} I_{i,j}\right) \tag{3.35}$$

$$Std_{x,y}^{l,r} = \frac{1}{\tau}\left(\sqrt{\sum_{i=x-\frac{w}{2}}^{x}\sum_{j=y-\frac{h}{2}}^{y+\frac{h}{2}}\left(I_{i,j} - Mean_{x,y}^{l,r}\right)^2} - \sqrt{\sum_{i=x}^{x+\frac{w}{2}}\sum_{j=y-\frac{h}{2}}^{y+\frac{h}{2}}\left(I_{i,j} - Mean_{x,y}^{l,r}\right)^2}\right) \tag{3.36}$$

$$Mean_{x,y}^{t,b} = \frac{1}{\tau}\left(\sum_{i=x-\frac{w}{2}}^{x+\frac{w}{2}}\sum_{j=y-\frac{h}{2}}^{y} I_{i,j} - \sum_{i=x-\frac{w}{2}}^{x+\frac{w}{2}}\sum_{j=y}^{y+\frac{h}{2}} I_{i,j}\right) \tag{3.37}$$

$$Std_{x,y}^{t,b} = \frac{1}{\tau}\left(\sqrt{\sum_{i=x-\frac{w}{2}}^{x+\frac{w}{2}}\sum_{j=y-\frac{h}{2}}^{y}\left(I_{i,j} - Mean_{x,y}^{t,b}\right)^2} - \sqrt{\sum_{i=x-\frac{w}{2}}^{x+\frac{w}{2}}\sum_{j=y}^{y+\frac{h}{2}}\left(I_{i,j} - Mean_{x,y}^{t,b}\right)^2}\right) \tag{3.38}$$

Since the chromaticity distribution is more relevant for color image classification, the above four features are calculated only for L channel. The patch based features $V_{x,y}^T$ for a pixel $I_{x,y}$ comprise of:

$$V_{x,y}^T = [V_{x,y}^I, (Mean_{x,y}, Std_{x,y}, Skw_{x,y})^{L,a,b}, \left(Mean_{x,y}^{l,r}, Std_{x,y}^{l,r}, Mean_{x,y}^{t,b}, Std_{x,y}^{t,b}\right)^L] \tag{3.39}$$

It is noted that $V_{x,y}^T$ also includes 6-D pixel based features $V_{x,y}^I$. One advantage of the extracted patch based features is that they are invariant to scale changes in images and retain high robustness against image rotations. This is crucial for roadside data analysis because real-world objects may be captured with different resolutions and rotations due to shakes of cameras mounted on a driving vehicle, and changes of the distances from cameras to objects. Thus, they are expected to work robustly on real-world data.

3.7.2.4 Pixel Probability Map (PPM)

For each of all pixels $I_{x,y}$ in an image I and a set of object categories C_M, the task of object classification is to generate a mapping function $\varphi : I_{x,y} \rightarrow C_M$, so that each image pixel is assigned to one object category. Based on the features extracted for $I_{x,y}$, this part adopts machine learning classifiers to obtain Pixel Probability Maps (PPMs), which represent the likelihoods of image pixels belonging to all classes. The PPMs are consequently used to create enhanced contextual superpixel probability maps.

Instead of building a traditional single multiple-class classifier that classifies all classes at a time, which may have difficulty distinguishing objects with similar visual appearance, we train a class-specific binary classifier separately for each of all classes, to more accurately reflect the probabilities of all pixels belonging to a specific object. Class-specific classifiers are able to focus on generating a mapping function from the extracted features to only one class and treating feature variations in the remainder classes as a second category. They are anticipated to be more effective in overcoming feature variations and improving classification performance. For multiple-class classification problems, class-specific classifiers produce a PPM for each of all classes.

We train a class-specific binary ANN classifier for each of all classes. However, other popular classifiers, such as SVM, can also be used here. Let C_i represent the ith class $(i = 1, 2, \ldots, M)$ and M be the total number of classes, the probability of a pixel $I_{x,y}$ characterized by the feature vector $V_{x,y}$ belonging to C_i can be predicted by the ith class-specific binary ANN classifier:

$$p^{i}_{x,y} = tran(w_i V_{x,y} + b_i) \tag{3.40}$$

where, $tran$ stands for a three-layer neural network with a tan-sigmoid activation function, w_i and b_i are trainable weights and constant parameters respectively for the ith class-specific classifier. A classifier generates a probability map for each class, and there are M maps in total. A pixel $I_{x,y}$ in an image is associated with a total of M probabilities, one for each class:

$$PPM_{x,y} = [p^{1}_{x,y}, p^{2}_{x,y}, \cdots, p^{M}_{x,y}] \tag{3.41}$$

3.7.3 Components of Approach

This section describes three main components of the SCSM approach, including PPS feature, contextual superpixel probability map, and superpixel growing.

3.7.3.1 Pixel Patch Selective (PPS) Feature

The PPS features are designed to handle the regional boundary problem. They adaptively select either pixel or patch based features for the classification of all pixels based on segmented superpixels. They are based on the observation that extracting features from a patch around the object boundaries unavoidably introduces a certain level of noise into the extracted feature set, while pixel based features generally do not suffer from this problem, but they cannot capture discriminative texture features. Thus, it is advisable to design a feature extraction method that can automatically choose to extract pixel or patch based features, depending on whether the current pixel is boundary or non-boundary.

A question arising is how to determine boundary versus non-boundary pixels? Fortunately, the segmented superpixels provide a clear partition of highly homogeneous regions and pixels forming their boundaries can be approximately regarded as boundary pixels. Given boundary coordinates of all superpixels and for a patch with a certain size (e.g. 7 × 7 pixels), a method is presented to classify all pixels into boundary versus non-boundary pixels as shown in Fig. 3.29. The non-boundary interior pixels are defined as being at least a half of the height (or width) of the patch far away from superpixel boundaries, where patch based features can be extracted accurately without introducing any noise. Pixels in image borders are determined in the same manner. The PPS features for a pixel $I_{x,y}$ are obtained by:

$$PPS_{x,y} = \begin{cases} V^I_{x,y} & \text{if} \quad I_{x,y} \in \text{boundary pixel} \\ V^T_{x,y} & \text{if} \quad I_{x,y} \in \text{non-boundary pixel} \end{cases} \tag{3.42}$$

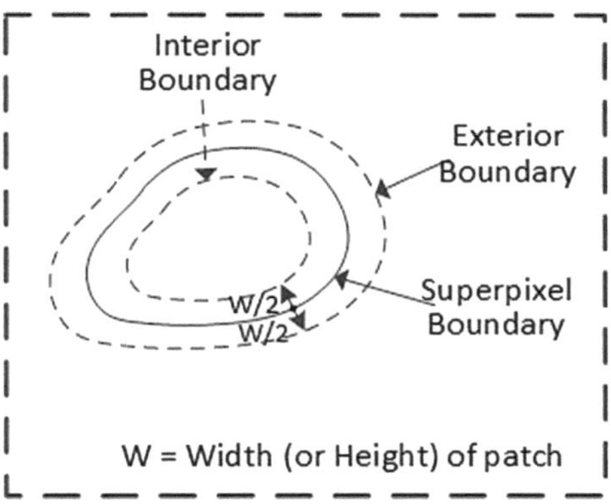

Fig. 3.29 Illustration of determining boundary versus non-boundary pixels. Given a segmented superpixel, the interior and exterior boundaries are determined by taking a distance of a half of the height (or width) of a patch away from the superpixel boundary. All pixels within the region between interior and exterior boundaries are treated as boundary pixels, while those within interior boundary are non-boundary pixels

3.7.3.2 Contextual Superpixel Probability Map (CSPM)

The Pixel Probability Maps (PPMs) for all classes are obtained using class-specific ANN classifiers based on PPS features. A drawback of ANN classifiers is that they treat each individual pixel separately without considering the contextual spatial information. In natural scenes, the geometric locations of objects may vary significantly from one image to another due to substantially varied viewpoints and scene content, making it difficult to learn statistical modelling (e.g. Gaussian and graphical models [54]) of spatial distribution of objects that works robustly for new test data. For roadside data analysis where roadside images are captured by a left camera mounted on a driving car, prior knowledge of locations of specific objects in the image can be utilized thanks to a fixed-viewpoint camera. As an instance, sky is unlikely to present at the bottom part of the image, while trees are unlikely to exist at the top part of the image. Thus, such contextual spatial information can be used to improve the PPMs.

Three simple contextual models are considered for three objects including road, sky, and tree respectively. Unlike current studies [55–58] that generate context-aware probabilistic spatial models based on inter-class relative spatial relationships, which completely discard absolute spatial coordinates of objects in images, the SCSM approach generates contextual models based on prior about the locations of object pixels in equally divided spatial blocks, which keep a trade-off between relative and absolute spatial relationships. Figure 3.30 illustrates the distribution of weight values $w_c, c \in \{\text{sky}, \text{road}, \text{tree}\}$, depending on the (x, y) coordinates of pixels in a test image. Equation (3.41) can be revised to:

$$PPM_{x,y} = [w_1 p^1_{x,y}, w_2 p^2_{x,y}, \ldots, w_M p^M_{x,y}] \tag{3.43}$$

where, w_i is the weight given to the ith class. Although the contextual models are simple, they are quite effective in correcting misclassification error in the SCSM.

All the above steps are performed at a pixel level, and we present CSPMs to perform classification at a superpixel level. The CSPMs are obtained by aggregating

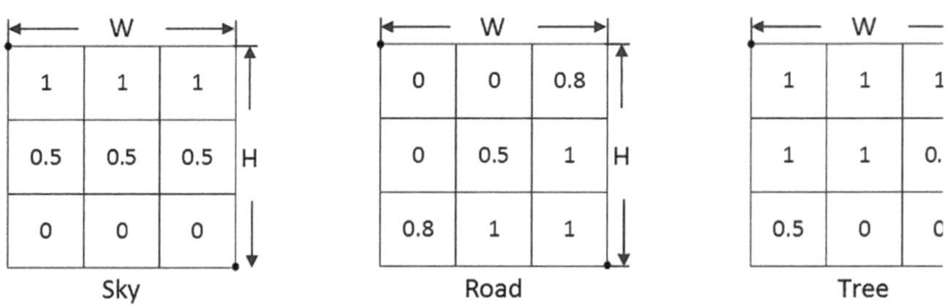

Fig. 3.30 Distribution of probability weights given to sky, road and tree, depending on the (x, y) coordinates of pixels in an image of Height × Width (H × W) pixels. The coordinates start from $(0, 0)$ at the *left-top corner* to the (H, W) at the *right-bottom corner*

probabilities over all pixels within each superpixel. For an input image I, it is segmented into N superpixels, i.e. $S = [S_j], j = 1, 2, \ldots, N$, using the graph-based algorithm [23] and the jth superpixel is S_j. For all pixels $I_{x,y} \in S_j$, the corresponding CSPM for the ith class is calculated using:

$$CSPM_j^i = \frac{1}{\tau_j} \sum_{I_{x,y} \in S_j} w_i p_{x,y}^i \tag{3.44}$$

where, τ_j is the number of pixels $I_{x,y}$ in S_j. Thus, the resulting CSPMs are composed of only one probability for each class and each superpixel:

$$CSPM_j = [CSPM_j^1, CSPM_j^2, \ldots, CSPM_j^M] \tag{3.45}$$

The above method of aggregating probabilities of all pixels within a superpixel for each class is analogous to performing decision-level majority voting over a pool of pixels, which plays an important role in removing classification error and improving the performance.

3.7.3.3 Contextual Superpixel Merging

This part introduces a contextual superpixel merging algorithm for superpixel classification, which utilizes both local spatial correlations between neighboring superpixels, and global contextual constraints between each superpixel and class labels. Unlike existing region merging methods which grow regions starting from the initial seeds with high confidence in class probabilities, the contextual superpixel merging algorithm grows superpixels in an opposite way, which selects superpixels with low probabilities in CSPMs and iteratively merges them into their closest neighbors based on two local and one global spatial constraints:

Local Constraint 1: *A neighbor Q of superpixel S_j accepts S_j when S_j has the highest similarity with Q among the similarities between Q and all Q's neighbors.*

Local Constraint 2: *A superpixel S_j accepts its neighbor Q when S_j has the highest similarity with Q among the similarities between S_j and all S_j' neighbors.*

Explanation: This condition enforces a local spatial constraint that only neighboring superpixels can be merged and they are the closest neighbor of each other. This double-side check prevents from merging an isolated superpixel that has a class label different from its neighbors.

Global Constraint: *A class category C accepts superpixel S_j when S_j has the highest similarity with C among the similarities between S_j and all classes.*

Explanation: This is common knowledge in machine learning, as superpixels from the same category should be close to each other in the feature space, while far away from those from other categories.

The superpixel merging algorithm comprises four steps:

(1) Calculate the similarity between superpixels. Histograms are one of the most widely adopted metrics for measuring the similarity between images. They collect statistical occurrence frequencies of features over an image, and thus are robust against noise and object variations. The SCSM approach represents the appearance in each superpixel using histograms of bag-of-words features. The visual dictionary of words is calculated by performing K-means clustering on pixel based features, i.e. $V^I = [R, G, B, L, a, b]$ in the training dataset. The pixel based features $V^I_{x,y}$ of a pixel $I_{x,y}$ in a superpixel S_j in a test image can then be quantized into one of K clustered words $W = [W^i], i = 1, 2, \ldots, K$ using the Euclidean distance:

$$I_{x,y} \in W^i \quad \text{if} \quad E(V^I_{x,y}, W^i) = \min_{i=1,2,\ldots,K}\left(\sum_{\gamma=1}^{\pi} \sqrt{(V^I_{x,y,\gamma} - W^i_{\gamma})^2}\right) \tag{3.46}$$

where, $V^I_{x,y,\gamma}$ is the γth element of $V^I_{x,y}$, W^i_{γ} is the γth element of W^i, $1 \leq \gamma \leq \pi$ and π is the length of elements in $V^I_{x,y}$. All pixels in S_j are then aggregated into K histogram bins:

$$\mathrm{H}^i_{S_j} = \sum_{\mathrm{I}_{x,y} \in \mathrm{S}_j} W^i \tag{3.47}$$

$$H_{S_j} = [H^1_{S_j}, H^2_{S_j}, \ldots, H^K_{S_j}] \tag{3.48}$$

where, $\mathrm{H}^i_{S_j}$ is the ith bin of the histogram H_{S_j}. The H_{S_j} is further normalized by summing all bins into a unit value of 1:

$$\hat{H}_{S_j} = [\hat{H}^1_{S_j}, \hat{H}^2_{S_j}, \ldots, \hat{H}^K_{S_j}] = [H^1_{S_j}/U, H^2_{S_j}/U, \ldots, H^K_{S_j}/U] \tag{3.49}$$

$$U = \sum_{i=1}^{K} H^i_{S_j} \tag{3.50}$$

For small resolutions of superpixels, the resulting histograms may be very sparse and contain only a handful of non-zero elements. Therefore, it is often advisable to set a small value to K. The similarity between two superpixels S_j and S_k can then be measured by calculating the Bhattacharyya coefficient $\mathrm{B}(S_j, S_k)$ of their normalized histograms $\hat{H}_{S_j}$ and $\hat{H}_{S_k}$:

$$B(S_j, S_k) = \sum_{i=1}^{K} \sqrt{\hat{H}^i_{S_j} * \hat{H}^i_{S_k}} \tag{3.51}$$

The higher $B(S_j, S_k)$ is, the higher the similarity between S_j and S_k is. Amongst existing statistical metrics such as χ^2 and Euclidean distances, the Bhattacharyya coefficient is chosen because it reflects perceptual similarity between regions and has shown good performance region merging [50].

(2) Select superpixel seeds. Two separate sets of superpixel seeds with low and high probabilities respectively in CSPMs are selected. The seeds with high probabilities are served as high confident regions which reflect local characteristics of objects in the test image, while those with low probabilities are superpixel candidates that should be merged into their neighbors. For the jth superpixel S_j, i.e. $S_j \in S$, its class probabilities in CSPMs for M classes can be calculated using (3.44), and the highest value among all classes indicates the class that S_j most likely belongs to:

$$p_j = \max_{i=1,\ldots,M} CSPM_j^i \tag{3.52}$$

A set of superpixel seeds with low confidence in class probabilities is selected from all superpixels by setting a threshold T to (3.52):

$$S_j \in Seed_i^l \quad \text{if} \quad p_j < T \text{ for class } C_i \tag{3.53}$$

A higher T indicates that more superpixels will be included in the merging process. The superpixels with the top P highest p_j for each class are selected as superpixel seeds with high confidence. If the number of superpixels for a class is less than P, only superpixels available are selected:

$$S_j \in Seed_i^h \quad \text{if} \quad p_j \in top\,P \text{ for class } C_i \tag{3.54}$$

(3) Merge local superpixels. The local superpixel merging algorithm merges superpixel seeds with low confidence from (3.53) into their most similar adjacent superpixels which have already been labelled to a class category. For this aim, many existing methods directly compare the similarities of each seed with all its neighbors, and then choose the neighbor with the highest similarity. These methods are essentially based on the information regarding which neighbor the seed wants to be merged into, but they may result in biased decisions as they do not consider larger contextual neighborhood information of all neighbors of the seed (e.g. are the neighbors of the seed willing to 'accept' it?). Thus, an alternative algorithm is adopted to perform double checks on whether the seed and its neighbors are willing to accept each other. This algorithm extends comparisons of the similarities between superpixels to a larger context so that it is anticipated to lead to more robust classification.

This algorithm is composed of two steps:

(a) Compare the similarities of all neighbors of a superpixel seed with their adjacent superpixels, to determine which neighbors are willing to 'accept' the seed. This process iteratively compares all neighbors of each seed with their adjacent superpixels and makes a decision on accepting or not based on whether the seed is the most similar superpixel among all adjacent superpixels of the seed's neighbors.

For a superpixel seed $S_j \in Seed_i^l$ for class C_i, let $\bar{M}_j = [M_v], v = 1, 2, \ldots, V$ be the set of adjacent superpixels of S_j and M_v be the vth member of $\bar{M}_j$. Obviously, M_v is an adjacent superpixel of S_j and let $\bar{Q}_{M_v} = [Q_\varphi], \varphi = 1, 2, \ldots, Q$ be the set of adjacent superpixels of M_v and Q_φ be the φth member of $\bar{Q}_{M_v}$. The similarity of M_v and Q_φ can be calculated using (3.51), i.e. $B(M_v, Q_\varphi)$. Because S_j is a member of $\bar{Q}_{M_v}$, and if the similarity $B(M_v, S_j)$ between M_v and S_j is the highest among all $B(M_v, Q_\varphi), \varphi = 1, 2, \ldots, Q$, then we label M_v as a neighbor which accepts S_j and add M_v into the set $\bar{A}_j$. It is completed using:

$$\bar{A}_j = \bar{A}_j \cup M_v \quad \text{if} \quad B(M_v, Q_t) = \max_{\varphi=1,2,..,Q} B(M_v, Q_\varphi) \quad \text{and} \quad Q_t = S_j \tag{3.55}$$

where, $\bar{A}_j$ stands for a set of neighbors of S_j that accepts S_j, and $1 \le t \le Q$. The 'max' operation performs iterative checks on all adjacent superpixels of M_v. Only when S_j has the maximum similarity with M_v among all adjacent superpixels of M_v, can it be determined that M_v is willing to accept S_j.

The above process repeats for all members in $\bar{M}_j$. The whole set of neighbors $\bar{A}_j$ accepting S_j is formed using:

$$\bar{A}_j = \bigcup_{v=1,2,\ldots,V} M_v \tag{3.56}$$

s.t. $M_v \in \bar{M}_j$ *and* M_v satisfies (3.55)

(b) Compare the similarities of each seed with all its accepting neighbors, to determine which neighbor is willing to be merged into. If there are multiple neighbors that are willing to accept the seed, only the neighbor that has highest similarity with the seed is merged into.

This process is completed by performing similarity comparisons between S_j and all members in $\bar{A}_j$, and selecting the neighbor A_r of S_j in $\bar{A}_j$ that has the highest similarity value:

$$A_r \text{ if } B(S_j, A_r) = \max_{l=1,2,..,L; A_l \in \bar{A}_j} B(S_j, A_l) \tag{3.57}$$

where, A_l is the lth member of $\bar{A}_j$, L the number of all members in $\bar{A}_j$, and $1 \le r \le L$. The 'max' operation sets a strict rule that the seed S_j is merged into only the most homogenous neighbor, and as a result, misclassification is introduced at a minimal level and contextual information between superpixels is taken into account in a larger neighorhood. When no neighbor wants to accept S_j, the label of S_j is determined using (3.52).

(4) Global superpixel refinement. The above unsupervised superpixel merging actually considers only the spatial correlation of a superpixel with its neighbors in a local neighborhood. It does not consider the global contextual information in the test image, which may vary substantially in different images. To reflect such contextual information in the test image, we use superpixel seeds with high confidence predicted using ANN classifiers in (3.54).

Let $S_u \in Seed_i^h, u = 1, 2, \ldots, P$ be the uth superpixel with a high confidence for the ith class C_i from (3.54), its K-bin histogram features are calculated using:

$$H_{S_u} = [\mathrm{H}_{S_u}^1, \mathrm{H}_{S_u}^2, \ldots, \mathrm{H}_{S_u}^K] \tag{3.58}$$

Aggregating each histogram bin over all P superpixels for C_i:

$$H_{C_i} = [\sum_{S_u \in Seed_i^h} \mathrm{H}_{S_u}^1, \sum_{S_u \in Seed_i^h} \mathrm{H}_{S_u}^2, \ldots, \sum_{S_u \in Seed_i^h} \mathrm{H}_{S_u}^K] \tag{3.59}$$

The global histogram features H_{C_i} of C_i can be converted into a normalized histogram $\hat{H}_{C_i}$ using (3.49) and (3.50). The similarity $B(S_j, \hat{H}_{C_i})$ between a superpixel seed $S_j \in Seed_i^l$ and $\hat{H}_{C_i}$ is then calculated using (3.51).

Assume that S_j is decided to be merged into A_r using (3.57), and A_r belongs to the zth class, i.e. $A_r \in C_z$ and $1 \le z \le M$, S_j is expected to have the highest similarity with $\hat{H}_{C_y}$ among the similarities between S_j and all classes $\hat{H}_{C_i}$:

$$A_r = A_r \cup S_j \quad \text{if} \quad B(S_j, \hat{H}_{C_z}) = \max_{i=1,2,..,M} B(S_j, \hat{H}_{C_i}) \tag{3.60}$$

The seed superpixel S_j is finally merged into A_r by assigning the same class category to C_z, i.e. $S_j \in C_z$.

The above superpixel refinement enforces the global constraint that a superpixel seed belonging to a class should have a higher similarity with superpixels in this class than with those in other classes.

The whole algorithm is summarized in Algorithm 3.1.

Algorithm 3.1 Spatial Constraint Superpixel Merging

Input: initial superpixel acquisition $S_j \in S$.
Output: all superpixels labeled to a class.

Step 1 - Initial seed selection:
Get $S_j \in Seed_i^l$ and $S_u \in Seed_i^h$ by setting T and P.
For each superpixel $S_j \in Seed_i^l$.

Step 2 – local superpixel merging:
For each adjacent superpixel of S_j, i.e. $M_v \in \bar{M}_j$.
For each adjacent superpixel of M_v, i.e. $Q_\varphi \in \bar{Q}_{M_v}$.
Calculate the similarity between M_v and Q_φ, i.e. $B(M_v, Q_\varphi)$.
End
Find the most similar adjacent superpixel of M_v:

$$Q_t \text{ if } B(M_v, Q_t) = \max_{\varphi=1,2,..,Q} B(M_v, Q_\varphi)$$

If $Q_t = S_j$, M_v accepts S_j and adds M_v to set $\bar{A}_j$;
Else M_v does not accept S_j.
End

For each adjacent superpixel accepting S_j, i.e. $A_l \in \bar{A}_j$.
Calculate similarity between S_j and A_l, i.e. $B(S_j, A_l)$.
End

Find the most similar adjacent superpixel of S_j:

$$A_r \text{ if } B(S_j, A_r) = \max_{l=1,2,..,L} B(S_j, A_l)$$

If $A_r \neq \emptyset$, S_j accepts A_r;
Else S_j remains the same label.

Step 3 – global superpixel refinement:
For each class $C_i \in \bar{C}$
Obtain the global features of C_i using $S_u \in Seed_i^h$:

$$H_{C_i} = [\sum_{S_u \in Seed_i^h} \mathrm{H}_{S_u}^1, \sum_{S_u \in Seed_i^h} \mathrm{H}_{S_u}^2, \cdots, \sum_{S_u \in Seed_i^h} \mathrm{H}_{S_u}^K]$$

Calculate the similarity between S_j and normalizedH_{C_i}.
End

Find the most similar class of S_j:

$$C_z \text{ if } B(S_j, \hat{H}_{C_z}) = \max_{i=1,2,..,M} B(S_j, \hat{H}_{C_i})$$

If $A_r \in C_z$, S_j is merged to A_r;
Else S_j remains the same label.
End

3.7.4 *Experimental Results*

The performance of the SCSM is evaluated on the cropped roadside object dataset, the natural roadside objet dataset, and the Stanford background benchmark dataset.

3.7.4.1 Implementation Details and Parameter Settings

The parameters of the graph-based algorithm are set based on the recommended setting in [24], i.e. $\sigma = 0.5, k = 80$, and $min = 80$ for an image size of 320×240 pixels. The class-specific ANN classifier is trained using the cropped roadside object dataset. To ensure equal training data for all classifiers, 80 pixels are selected from random locations in each cropped region. The ANNs have three layers and are trained using the Levenberg-Marquardt backpropagation algorithm, i.e. a goal error of 0.001, 500 maximum epochs, and a learning rate of 0.01. The input layer has 6 and 19 neurons respectively for pixel based and patch based features. The size of patches is set to 7×7 pixels. The K for K-means clustering is set to 40. The program was developed using the Matlab on a laptop with 4 GB memory and 2.4 GHz CPU.

Evaluation metrics: The performance of the SCSM approach is measured using two metrics: overall accuracy measured in terms of all pixels across all test images and class accuracy averaged over all object classes using pixelwise comparisons between classified results and ground truths. Four-fold random cross validations are used, and the average accuracy over all validations is used as the results. In each validation, 75% of the cropped regions of each class are randomly selected for training and the rest 25% for testing.

3.7.4.2 Classification Results on Cropped Roadside Object Dataset

Figure 3.31 and Table 3.26 show the class accuracy obtained using class-specific ANN classifiers for seven objects using pixel based versus patch based features. The results are obtained based on pixel-level classification without superpixel merging. The classifiers produce more than 90% accuracies for all objects for both pixel and patch based features, except for brown grasses whose accuracy is close to 90%. Patch based features have around 2% higher accuracies than pixel based features for all classes, and this confirms the benefit of using texture features in spatial local patches for more accurate classification. However, the rank of all classes in terms of accuracy is little impacted by the use of pixel or patch based features, indicating consistent intrinsic difficulty in classifying all classes. Sky has more than 99% accuracy and is the easiest object for correct classification, while brown grass and soil are the most difficult classes. Using a different number of hidden neurons has little impact on the results, and using more hidden neurons only improves the performance marginally.

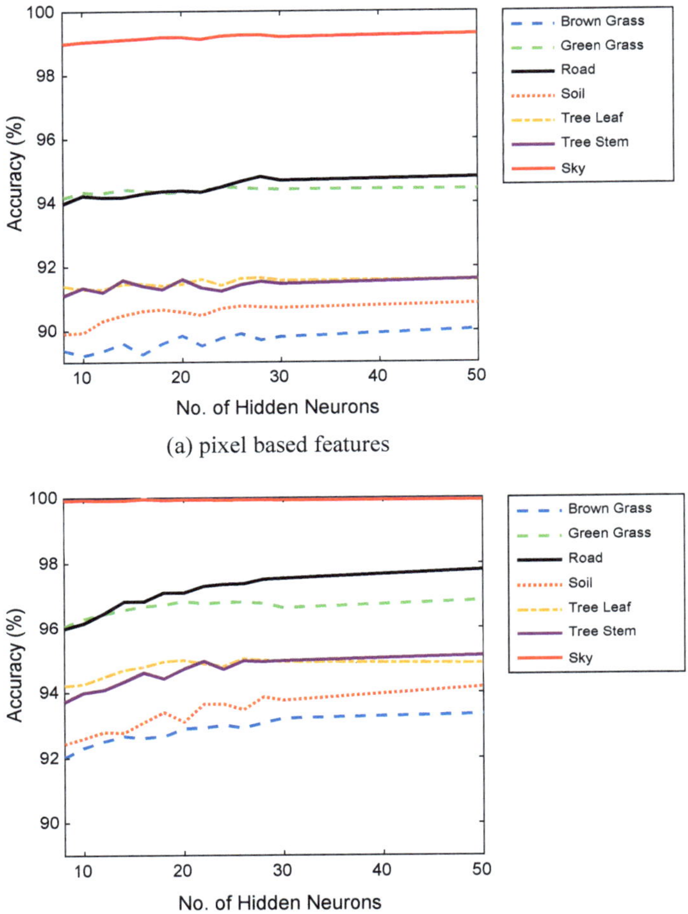

(a) pixel based features

(b) patch based features

Fig. 3.31 Classification accuracy (%) versus number of hidden neurons on the cropped roadside object dataset. Patch based features outperform pixel based features for all objects

3.7.4.3 Classification Results on Natural Roadside Object Dataset

Figure 3.32 shows the overall accuracy on the natural roadside object dataset. Seven approaches are included for performance comparisons, including pixel based features with or without spatial constraints (i.e. Pixel-C and Pixel-NC), patch based features with or without spatial constraints (i.e. Patch-C and Patch-NC), PPS features with or without spatial constraints (i.e. PPS-C and PPS-NC), and the SCSM model. From the figure, we can observe that considering the spatial constraints leads to big increases to the overall accuracy for all three types of features,

Table 3.26 Class accuracy (%, ± standard deviation) of pixel versus patch based features on the cropped roadside object dataset

Feature	Brown grass	Green grass	Road	Soil	Tree leaf	Tree stem	Sky
Patch	93.0 ± 0.2	96.8 ± 0.2	97.4 ± 0.2	93.7 ± 0.2	94.8 ± 0.1	94.7 ± 0.3	99.9 ± 0.0
Pixel	89.8 ± 0.2	94.5 ± 0.1	94.5 ± 0.1	90.7 ± 0.3	91.4 ± 0.3	91.3 ± 0.4	99.2 ± 0.1

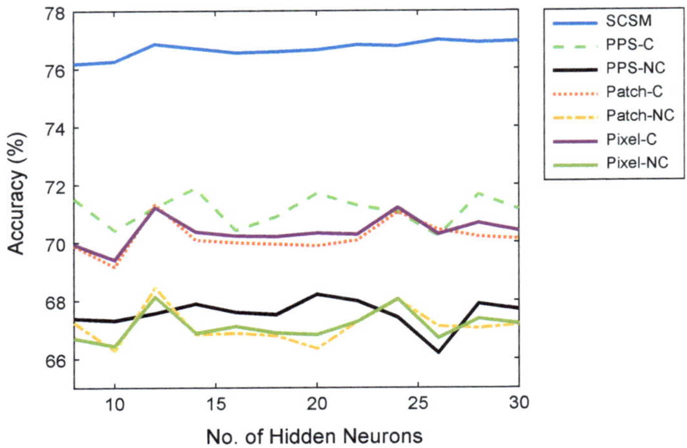

Fig. 3.32 Overall accuracy comparisons between approaches on the natural roadside object dataset. The SCSM outperforms all benchmark approaches. Patch based features show higher accuracies than both PPS and pixel based features for all classes

including pixel based, patch based and PPS features. For both the cases of using or not using spatial constraints, PPS features have higher overall accuracy than patch based and pixel based features. This confirms our anticipation that PPS features take into account the noise introduced at regional boundaries between superpixels. Unlike the results on the cropped dataset, pixel based features slightly outperform patch based features, probably because the latter gets more confusion in classifying objects in boundary regions. The SCSM approach significantly outperforms all benchmark approaches, confirming the importance of integrating local and global contextual constraints in the superpixel merging process.

We also investigated the sensitivity of the SCSM approach to threshold values T in (3.53) and P in (3.54). T controls the number of superpixels selected as having low confidence and need to be merged. If T decreases to 0, no superpixel is merged, and if T equals to 1, all superpixels are processed by the merging algorithm. P controls those superpixels with high confidence, which represent global constraints of all classes. Our results show that $T = 0.5$ produces the best result while merging less or more superpixels has little impact on the results. The classification results are nearly the same for using $P = [1, 3, 5, 7]$, indicating that the number of top superpixels used for each class has little impact on the performance of the SCSM approach.

Table 3.27 shows the class accuracy of the SCSM approach. Similar to the results on the cropped dataset, sky is the easiest object for classification with 97.4% accuracy, while soil is the most difficult object with only 50.2% accuracy. A significant proportion (41.7%) of soil pixels are misclassified as brown grass, probably because they share a similar yellow color. In addition, we also observed

Table 3.27 Confusion matrix of six classes obtained using the SCSM on the natural roadside object dataset

	Brown grass	Green grass	Road	Soil	Tree	Sky
Brown grass	**74.5**	13.8	2.0	5.5	4.2	0.0
Green grass	10.2	**79.0**	2.8	0.7	7.3	0.0
Road	8.4	0.5	**78.0**	12.7	0.4	0.0
Soil	41.7	6.0	1.5	**50.2**	0.6	0.0
Tree	6.3	6.0	2.4	0.7	**79.8**	4.8
Sky	0.2	0.0	1.1	0.3	0.9	**97.4**

Overall accuracy = 77.4%, Hidden neuron no. = 26, $T = 0.5$ and $P = 5$
The bold figure indicates class accuracy for each object

that lighting variations lead to confusion between soil and brown grass pixels. Brown and green grasses are prone to be misclassified to each other as well, as discriminating between them is even hard for human eyes. Road also tends to be misclassified as soil.

Figure 3.33 compares the classification results of a set of samples using different types of features. Overall, PPS features have similar results to pixel based features, but they are able to correct some classification errors. The SCSM approach produces the smoothest and most accurate results by enforcing spatial constraints at a superxpiel level. We also analyze failure cases to provide useful insights into challenging factors for natural roadside object classification, which are important for further improvements to the SCSM approach. Our analysis found that there is a substantial portion of misclassification between brown grasses and soil (also trees) and shadows of objects also result in notable misclassification of brown grass pixels into road and trees. The images demonstrate significant impact of color similarities between objects and varying lighting conditions on the results. Therefore, they should receive special attention for roadside vegetation classification in real-world conditions.

3.7.4.4 Classification Results on Stanford Background Data

Although the SCSM model is designed for roadside object classification, it can be easily modified for generic object categorization in various scene data. This can be achieved by simply ignoring the three contextual models used specifically for road, sky, and tree in roadside images, and setting an equal probability weight for all objects in calculating CSPMs.

Table 3.28 compares the performance of the SCSM with state-of-the-art methods on the Stanford background dataset. Following the commonly used evaluation procedure [59], 5-fold cross validations are conducted. In each cross validation, 572 images are randomly selected for training and the rest 143 images for testing. For fair comparisons, the same set of over-segmented superpixels, and same superpixel-level color, texture and geometric features [59] are used in the experiments. We can see that the SCSM has higher classification accuracy than the approach [59] and produces accuracy comparable to the state-of-the-art methods.

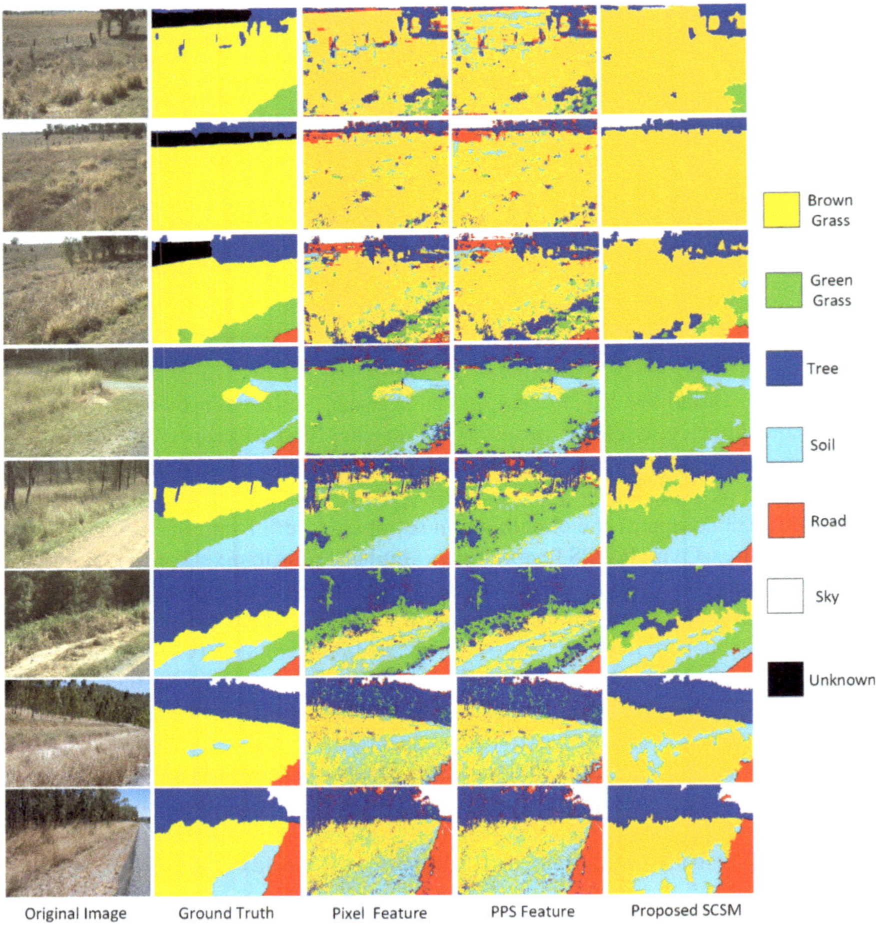

Fig. 3.33 Comparisons of classification results of the SCSM with using pixel based features and PPS features. The SCSM produces more consistent results than using pixel or PPS features

Table 3.28 Performance (%) comparisons with state-of-the-art methods on the stanford background dataset

Ref.	Overall Acc.	Class Acc.
Gould et al. [59]	76.4	–
Munoz et al. [60]	76.9	66.2
Tighe et al. [61]	77.5	–
Socher et al. [62]	78.1	–
Kumar et al. [63]	79.4	–
Lempitsky et al. [64]	81.9	72.4
Farabet et al. [65]	81.4	76.0
SCSM	77.5	68.8

Table 3.29 Confusion matrix of eight classes on the stanford background dataset

	Sky	Tree	Road	Grass	Water	Building	Mountain	Foreground
Sky	**94.0**	2.5	0.1	0.0	0.3	2.1	0.5	0.5
Tree	5.2	**69.8**	1.0	3.5	0.3	14.8	1.8	3.6
Road	0.2	0.6	**89.3**	0.9	2.6	2.2	0.4	3.8
Grass	0.3	6.3	4.0	**81.0**	2.3	1.9	2.7	1.5
Water	2.4	0.8	21.8	3.1	**59.1**	4.2	3.9	4.7
Building	2.7	6.0	3.3	0.7	0.5	**78.7**	1.1	7.0
Mountain	6.1	20.4	7.0	7.2	3.9	25.1	**23.0**	7.3
Foreground	2.5	5.2	11.2	2.3	1.8	20.4	1.1	**55.5**

Hidden neuron no. = 16, $T = 0.35$, and $P = 5$
The bold figure indicates class accuracy for each object

The confusion matrix of eight classes in Table 3.29 indicates that mountain and water are the two most difficult objects for correct classification, while sky is the easiest one with 94% accuracy. The results agree with those in previous studies [59, 60], where sky and mountain have the highest and lowest accuracies of 92 and 14% respectively. Due to the similarity in color and texture, mountain and water are more likely to be misclassified as building and road respectively.

3.7.5 *Summary*

This section has presented a Spatial Contextual Superpixel Model (SCSM) for object classification in natural roadside images. The PPS features are introduced to handle noise in patch based feature extraction at regional boundaries. Contextual class probability maps are obtained by fusing class-specific ANN classifiers and contextual models for pixel-level object classification, and the classification results are further aggregated over superpixels. A superpixel merging strategy is then adopted to refine the results by merging superpixels with low confidence into their most similar neighbors. Experimental results indicate that the SCSM achieves over 90, 77.4, and 77.5% accuracies on the cropped roadside object, natural roadside object and Stanford background datasets respectively. The SCSM outperforms pixel or patch based features, confirming the benefit of considering local and global spatial contextual information at a superpixel level for object classification.

The SCSM still can be extended from several aspects. (1) The color and texture features used for ANN classification have not taken into account their different contributions. It is worth adopting a feature selection process to select a subset of most important features for each object to create class-specific features. (2) It is still possible to further improve the performance by using other classifiers such as SVM and ensemble classifiers rather than ANN. (3) The similarity metric is calculated using histograms of color features only, and thus one possible extension is adding texture features, such as filter-bank based texton features [15].

References

1. W.S. McCulloch, W. Pitts, A logical calculus of the ideas immanent in nervous activity. Bull. Math. Biophys. **5**, 115–133 (1943)
2. J. Schmidhuber, Deep learning in neural networks: an overview. Neural Networks **61**, 85–117 (2015)
3. L. Zhang, B. Verma, D. Stockwell, Roadside vegetation classification using color intensity and moments, in *the 11th International Conference on Natural Computation*, 2015, pp. 1250–1255
4. N.W. Campbell, B.T. Thomas, T. Troscianko, Automatic segmentation and classification of outdoor images using neural networks. Int. J. Neural Syst. **08**, 137–144 (1997)
5. D.V. Nguyen, L. Kuhnert, K.D. Kuhnert, Spreading algorithm for efficient vegetation detection in cluttered outdoor environments. Robot. Auton. Syst. **60**, 1498–1507 (2012)
6. K.E.A. Van De Sande, T. Gevers, C.G.M. Snoek, Evaluating color descriptors for object and scene recognition. IEEE Trans. Pattern Anal. Mach. Intell. **32**, 1582–1596 (2010)
7. F. Mindru, T. Tuytelaars, L.V. Gool, T. Moons, Moment invariants for recognition under changing viewpoint and illumination. Comput. Vis. Image Underst. **94**, 3–27 (2004)
8. C.C. Chang, C.J. Lin, Libsvm: a library for support vector machines, 2001. Software available at http://www.csie.ntu.edu.tw/cjlin/libsvm, 2001
9. C.J.C. Burges, A tutorial on support vector machines for pattern recognition. Data Min. Knowl. Disc. **2**, 121–167 (1998)
10. Z. Qi, Y. Tian, Y. Shi, Robust twin support vector machine for pattern classification. Pattern Recogn. **46**, 305–316 (2013)
11. S. Chowdhury, B. Verma, M. Tom, M. Zhang, Pixel characteristics based feature extraction approach for roadside object detection, in *International Joint Conference on Neural Networks (IJCNN)*, 2015, pp. 1–8
12. P. Jansen, W. Van Der Mark, J.C. Van Den Heuvel, F.C.A. Groen, Colour based off-road environment and terrain type classification, in *Intelligent Transportation Systems*, 2005, pp. 216–221
13. J. Malik, S. Belongie, T. Leung, J. Shi, Contour and texture analysis for image segmentation. Int. J. Comput. Vis. **43**, 7–27 (2001)
14. M.R. Blas, M. Agrawal, A. Sundaresan, K. Konolige, Fast color/texture segmentation for outdoor robots, in *IEEE/RSJ International Conference on Intelligent Robots and Systems (IROS)*, 2008, pp. 4078–4085
15. J. Winn, A. Criminisi, T. Minka, Object categorization by learned universal visual dictionary, in *Tenth IEEE International Conference on Computer Vision (ICCV)*, 2005, pp. 1800–1807
16. J. Shotton, J. Winn, C. Rother, A. Criminisi, Textonboost for image understanding: multi-class object recognition and segmentation by jointly modeling texture, layout, and context. Int. J. Comput. Vis. **81**, 2–23 (2009)
17. J. Shotton, M. Johnson, R. Cipolla, Semantic texton forests for image categorization and segmentation, in *IEEE Conference on Computer Vision and Pattern Recognition (CVPR)*, 2008, pp. 1–8
18. L. Zhang, B. Verma, D. Stockwell, Class-semantic color-texture textons for vegetation classification, in *Neural Information Processing*, 2015, pp. 354–362
19. Z. Haibing, L. Shirong, Z. Chaoliang, Outdoor scene understanding using Sevi-Bovw model, in *International Joint Conference on Neural Networks (IJCNN)*, 2014, pp. 2986–2990
20. D. Comaniciu, P. Meer, Mean shift: a robust approach toward feature space analysis. IEEE Trans. Pattern Anal. Mach. Intell. **24**, 603–619 (2002)
21. D. Yining, B.S. Manjunath, Unsupervised segmentation of color-texture regions in images and video. IEEE Trans. Pattern Anal. Mach. Intell. **23**, 800–810 (2001)
22. R. Xiaofeng, J. Malik, Learning a classification model for segmentation, in *Ninth IEEE International Conference on Computer Vision (ICCV)*, 2003, pp. 10–17

23. P. Felzenszwalb, D. Huttenlocher, Efficient graph-based image segmentation. Int. J. Comput. Vis. **59**, 167–181 (2004)
24. C. Chang, A. Koschan, C. Chung-Hao, D.L. Page, M.A. Abidi, Outdoor scene image segmentation based on background recognition and perceptual organization. IEEE Trans. Image Process. **21**, 1007–1019 (2012)
25. A. Bosch, X. Muñoz, J. Freixenet, Segmentation and description of natural outdoor scenes. Image Vis. Comput. **25**, 727–740 (2007)
26. Y. Kang, K. Yamaguchi, T. Naito, Y. Ninomiya, Multiband image segmentation and object recognition for understanding road scenes. IEEE Trans. Intell. Trans. Syst. **12**, 1423–1433 (2011)
27. Y. Lecun, L. Bottou, Y. Bengio, P. Haffner, Gradient-based learning applied to document recognition. Proc. IEEE **86**, 2278–2324 (1998)
28. L. Zheng, Y. Zhao, S. Wang, J. Wang, Q. Tian, Good practice in CNN feature transfer, *arXiv preprint* arXiv:1604.00133, 2016
29. I. Harbas, M. Subasic, CWT-based detection of roadside vegetation aided by motion estimation, in *5th European Workshop on Visual Information Processing (EUVIP)*, 2014, pp. 1–6
30. I. Harbas, M. Subasic, Motion Estimation Aided Detection of Roadside Vegetation, in *7th International Congress on Image and Signal Processing (CISP)*, 2014, pp. 420–425
31. I. Harbas, M. Subasic, Detection of roadside vegetation using features from the visible spectrum, in *37th International Convention on Information and Communication Technology, Electronics and Microelectronics (MIPRO)*, 2014, pp. 1204–1209
32. V. Balali, M. Golparvar-Fard, Segmentation and recognition of roadway assets from car-mounted camera video streams using a scalable non-parametric image parsing method. Autom. Constr. Part A **49**, 27–39 (2015)
33. B. Sowmya, B. Sheela Rani, Colour image segmentation using fuzzy clustering techniques and competitive neural network. Appl. Soft Comput. **11**, 3170–3178 (2011)
34. T. Kinattukara, B. Verma, Wavelet based fuzzy clustering technique for the extraction of road objects, in *IEEE International Conference on Fuzzy Systems (FUZZ)*, 2015, pp. 1–7
35. J. Schoukens, R. Pintelon, H.V. Hamme, The interpolated fast fourier transform: a comparative study. IEEE Trans. Instrum. Meas. **41**, 226–232 (1992)
36. M. Lotfi, A. Solimani, A. Dargazany, H. Afzal, M. Bandarabadi, Combining wavelet transforms and neural networks for image classification, in *41st Southeastern Symposium on System Theory*, 2009, pp. 44–48
37. T. Kinattukara, B. Verma, Clustering based neural network approach for classification of road images, in *International Conference on Soft Computing and Pattern Recognition (SoCPaR)*, 2013, pp. 172–177
38. T. Kinattukara, B. Verma, A neural ensemble approach for segmentation and classification of road images, in *Neural Information Processing*, 2014, pp. 183–193
39. A. Schepelmann, R.E. Hudson, F.L. Merat, R.D. Quinn, Visual segmentation of lawn grass for a mobile robotic lawnmower, in *IEEE/RSJ International Conference on Intelligent Robots and Systems (IROS)*, 2010, pp. 734–739
40. P. Kamavisdar, S. Saluja, S. Agrawal, A survey on image classification approaches and techniques. Int. J. Adv. Res. Comput. Commun. Eng. **2**, 1005–1009 (2013)
41. T.-H. Cho, R.W. Conners, P.A. Araman, A comparison of rule-based, K Nearest neighbor, and neural net classifiers for automated industrial inspection, in *The IEEE/ACM International Conference on Developing and Managing Expert System Programs*, 1991, pp. 202–209
42. M. Liu, Fingerprint classification based on adaboost learning from singularity features. Pattern Recogn. **43**, 1062–1070 (2010)
43. J. Petrová, H. Moravec, P. Slavıková, M. Mudrová, A. Procházka, Neural network in object classification using Matlab. Network **12**(10) (2012)
44. H.-Y. Yang, X.-Y. Wang, Q.-Y. Wang, X.-J. Zhang, LS-SVM based image segmentation using color and texture information. J. Vis. Commun. Image Represent. **23**, 1095–1112 (2012)

45. A. Rehman, Y. Gao, J. Wang, Z. Wang, Image classification based on complex wavelet structural similarity. Sig. Process. Image Commun. **28**, 984–992 (2012)
46. T.S. Hai, N.T. Thuy, Image classification using support vector machine and artificial neural network. Int. J. Inf. Technol. Comput. Sci. (IJITCS) **4**, 32 (2012)
47. S. Kang, S. Park, A fusion neural network classifier for image classification. Pattern Recogn. Lett. **30**, 789–793 (2009)
48. W.-T. Wong, S.-H. Hsu, Application of SVM and ANN for image retrieval. Eur. J. Oper. Res. **173**, 938–950 (2006)
49. S. Chowdhury, B. Verma, D. Stockwell, A novel texture feature based multiple classifier technique for roadside vegetation classification. Expert Syst. Appl. **42**, 5047–5055 (2015)
50. J. Ning, L. Zhang, D. Zhang, C. Wu, Interactive image segmentation by maximal similarity based region merging. Pattern Recogn. **43**, 445–456 (2010)
51. L. Zhang, B. Verma, D. Stockwell, Spatial contextual superpixel model for natural roadside vegetation classification. Pattern Recogn. **60**, 444–457 (2016)
52. P. Arbelaez, M. Maire, C. Fowlkes, J. Malik, Contour detection and hierarchical image segmentation. IEEE Trans. Pattern Anal. Mach. Intell. **33**, 898–916 (2011)
53. Y. Hui, L. Mingjing, Z. Hong-Jiang, F. Jufu, Color texture moments for content-based image retrieval, in *International Conference on Image Processing (ICIP)*, 2002, pp. 929–932
54. C. Myung Jin, A. Torralba, A.S. Willsky, A tree-based context model for object recognition. IEEE Trans. Pattern Anal. Mach. Intell. **34**, 240–252 (2012)
55. A. Singhal, L. Jiebo, and Z. Weiyu, "Probabilistic Spatial Context Models for Scene Content Understanding," in *Computer Vision and Pattern Recognition, (CVPR), IEEE Conference on*, 2003, pp. 235–241
56. Y. Jimei, B. Price, S. Cohen, Y. Ming-Hsuan, Context driven scene parsing with attention to rare classes, in *IEEE Conference on Computer Vision and Pattern Recognition (CVPR)*, 2014, pp. 3294–3301
57. S. Gould, J. Rodgers, D. Cohen, G. Elidan, D. Koller, Multi-class segmentation with relative location prior. Int. J. Comput. Vision **80**, 300–316 (2008)
58. B. Micusik, J. Kosecka, Semantic segmentation of street scenes by superpixel co-occurrence and 3D geometry, in *IEEE 12th International Conference on Computer Vision Workshops (ICCV Workshops)*, 2009, pp. 625–632
59. S. Gould, R. Fulton, D. Koller, Decomposing a scene into geometric and semantically consistent regions, in *IEEE 12th International Conference on Computer Vision (ICCV)*, 2009, pp. 1–8
60. D. Munoz, J.A. Bagnell, M. Hebert, stacked hierarchical labeling, in *European Conference on Computer Vision (ECCV)*, 2010, pp. 57–70
61. J. Tighe, S. Lazebnik, Superparsing: scalable nonparametric image parsing with superpixels, in *European Conference on Computer Vision (ECCV)*, 2010, pp. 352–365
62. R. Socher, C.C. Lin, C. Manning, A.Y. Ng, Parsing natural scenes and natural language with recursive neural networks, in *Proceedings of the 28th International Conference on Machine Learning (ICML)*, 2011, pp. 129–136
63. M.P. Kumar, D. Koller, Efficiently selecting regions for scene understanding, in *IEEE Conference on Computer Vision and Pattern Recognition (CVPR)*, 2010, pp. 3217–3224
64. V. Lempitsky, A. Vedaldi, A. Zisserman, Pylon model for semantic segmentation, in *Advances in Neural Information Processing Systems*, 2011, pp. 1485–1493
65. C. Farabet, C. Couprie, L. Najman, Y. LeCun, Learning hierarchical features for scene labeling. IEEE Trans. Pattern Anal. Mach. Intell. **35**, 1915–1929 (2013)

Chapter 4
Deep Learning Techniques for Roadside Video Data Analysis

In this chapter, we describe deep learning techniques that are proposed for roadside video data analysis. We firstly present an introduction to deep learning concepts, and a short review of several typical types of CNN. We then present an experimental investigation to compare the benefits of using features automatically extracted by CNN compared to using traditional hand engineering features. We also show a comparison of an ensemble of multiple CNNs with a single CNN or MLP classifier. Finally, we propose a deep learning architecture for roadside data analysis, and demonstrate its state-of-the-art performance on benchmark datasets compared with existing approaches.

4.1 Introduction

Deep learning techniques have recently gained increasing popularity and shown state-of-the-art performance in various computer vision tasks [1]. The main reason for this popularity is that they can learn discriminative and compact feature representations automatically from raw image pixels rather than using hand-engineered features and classify the learnt features into different object categories. Deep learning techniques are capable of inherently encoding discriminative features of each object while taking into account the variations within the object category or arising from the environment. Unlike traditional machine learning algorithms, there is no need for manual feature engineering such as feature extraction and selection in deep learning.

There are various types of deep learning techniques that have been proposed previously, such as CNN, recursive neural networks, deep belief networks, deep Boltzmann machines, and stacked auto-encoders. Although the actual structures of these techniques might vary significantly, they share a similar conceptual framework comprising of multiple layers of linear or non-linear processing units which progressively generate feature representations in each of all layers from low to high levels. Thus, more abstract and discriminative patterns can be obtained in the higher layers.

B. Verma et al., *Roadside Video Data Analysis: Deep Learning*,
Studies in Computational Intelligence 711, DOI 10.1007/978-981-10-4539-4_4

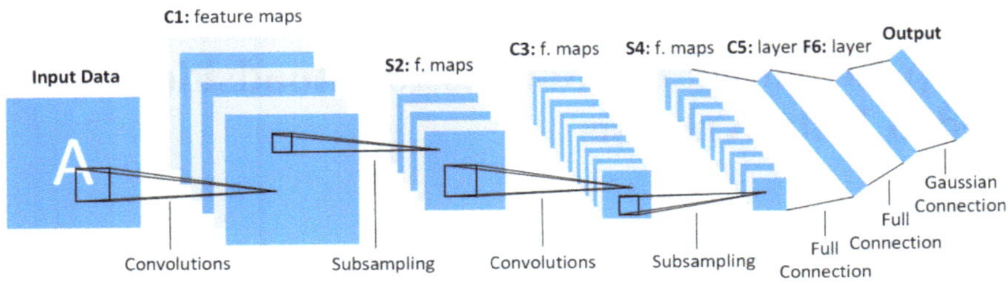

Fig. 4.1 Architecture of the LeNet-5 CNN. There are 7 layers in total excluding the input layer. The abstract feature representations are extracted by progressively applying convolutional and subsampling operations. The features are then fed into two full connection layers for classifying the input data into different object categories

CNN is one of the most widely known deep learning techniques for object classification. The design of CNN is motivated by the animal's visual cortex, i.e., the arrangement of cells and its learning process. It can also be considered as a biologically-inspired variant of MLP. The learning in CNN involves multiple processing layers, comprising multiple linear and non-linear transformations. Figure 4.1 shows the overall structure of the popular LeNet-5 CNN originally proposed for digit character recognition. The main processing layers include [INPUT—CONV—SUMP—POOL—FC] i.e., the input layer is fed to a CONVolutional (CONV) layer where a set of learnable filters are used. These filters are then fed to the REctified Linear Units (RELUs) to increase the non-linearity of the decision function. After the RELUs, a SUbsaMPling layer (SUMP) is used where non-linear downsampling is done, and finally a Fully Connected (FC) layer is used for the classification of objects [2].

4.2 Related Work

Over the past few years, the popularity of research on deep learning techniques has grown substantially. Figure 4.2 shows that the number of research articles on CNN from 1970 to 2016, indicating a substantial increase in research efforts on the investigation of CNN and its applications in various tasks.

Various types of CNN based deep learning architectures have been proposed for object classification tasks. Table 4.1 presents several typical CNN architectures and their reported applications.

Apart from the LeNet-5, Ba et al. [3] proposed a visual attention model based on deep recurrent neural networks and used reinforcement learning to train the model by attending to the most relevant regions of an input image. The model was evaluated on the MNIST dataset and a multi-digit street view house number dataset. The model was found to be more successful for house number recognition compared with CNN. Donahue et al. [4] proposed a recurrent convolutional

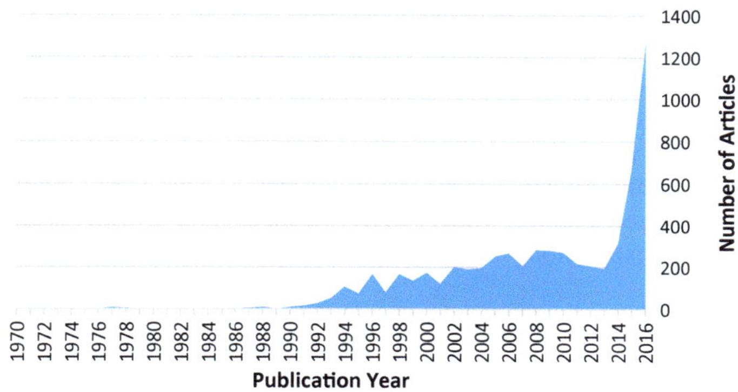

Fig. 4.2 CNN research trend from 1970 to 2016. *Source* Scopus

Table 4.1 A brief review of typical types of CNNs

Refs.	CNN type	Brief description of architecture	Application
LeCun et al. [13]	LeNet	First application of CNN to digit recognition. [INPUT—CONV—SUMP—CONV—SUMP—FC]	Zip codes, handwritten digits
Krizhevsky et al. [46]	AlexNet	Popularize CNN for computer vision. [CONV—5xMAX SUMP—FC]	Handwritten digits, ILSVRC 2010
Zeiler et al. [47]	ZF Net	Similar to AlexNet. [UNPOOLED MAPS—RECTIFIED—RECONSTRUCTION—POOL—RECTIFIED—FC]	ImageNet 2012, Caltech 101, Caltech 256, PASCAL 2012
Szegedy et al. [48]	GoogLeNet	Introduce an inception module to reduce parameters. [INPUT—CONV—POOL—INCEPTION—RELU—SOFTMAX]	ILSVRC 2012–2014
Simonyan et al. [49]	VGGNet	A similar configuration to GoogleNet without inception. [INPUT—3xMAXPOOL—3xFC—SOFTMAX]	ILSVRC 2012–2014
Kaiming et al. [50]	ResNet	Skip connection but use heavy batch normalization. No FC at the end	CIFAR 10, ILSVRC 2012

architecture suitable for large-scale visual learning which demonstrated promising results on the task of categorizing over 12,000 videos into 101 human action classes. Dundar et al. [5] proposed a clustering algorithm to reduce the number of correlated parameters for an increased accuracy. A patch feature extraction method was used to reduce the redundancy between filters at neighbouring locations. An accuracy of 74.1% was obtained on the STL-10 image recognition dataset and a test error of 0.5% on the MNIST dataset. Krizhevsky et al. [6] proposed the ImageNet deep convolutional neural network to classify over 1.2 million high-resolution images. The ImageNet has 60 million parameters and 650,000 neurons, consisting of five convolutional layers and three fully-connected layers with a final 1000-way softmax layer. Non-saturated neurons were used to ensure fast training along with an efficient GPU implementation. In [7], deep learning has been used for robotic hand grasp detection. A two-step cascade system with two deep networks was used, where the top detections from the first network were re-evaluated by the second network. A deep network was applied into black box image classification with 130,000 extra unlabelled samples [8]. In [9] a robust 4-layer CNN architecture was proposed for face recognition, which can handle facial images with occlusions, pose variations, facial expressions and varying illumination.

4.3 Automatic Versus Manual Feature Extraction

4.3.1 Introduction

Although CNN has been applied successfully to many computer vision tasks, it is still important to understand the advantage of the learning process in CNN over other existing techniques. The complexity of CNN makes it difficult to be directly used for some handy and small scale image processing tasks, where adequate training data is hard to obtain. CNN is well-known for its capacity of automatically extracting features from raw images. However, it is still unclear whether feature extraction incorporated in the deep learning architecture is any better than manual feature extraction techniques. There has been little research to systematically evaluate automatic feature extraction and classification abilities of the deep learning architecture. Therefore, it is important to conduct a systematic study to answer this question.

In this section, we conduct systematic experiments to compare CNN with the traditional MLP in image classification tasks [10], to answer the following questions: (1) is it always better to use CNN with automatic feature extraction for image classification? (2) how does CNN perform on non-complex datasets in comparison with the traditional MLP? and (3) how the performance of CNN can be further improved? We use MLP as a benchmark classifier, due to two reasons: (1) the classification layer after feature extraction in CNN can be represented as a MLP, and (2) MLP is a popularly used ANN which supports object classification with or without manual feature extraction.

4.3.2 Comparison Framework

Figure 4.3 presents an overview of the comparison framework [10], which is used to compare automatic feature extraction in CNN with manual feature extraction. The traditional MLP is used as the benchmark classifier for all models used for comparisons. The framework takes an input image and feeds it into three individual

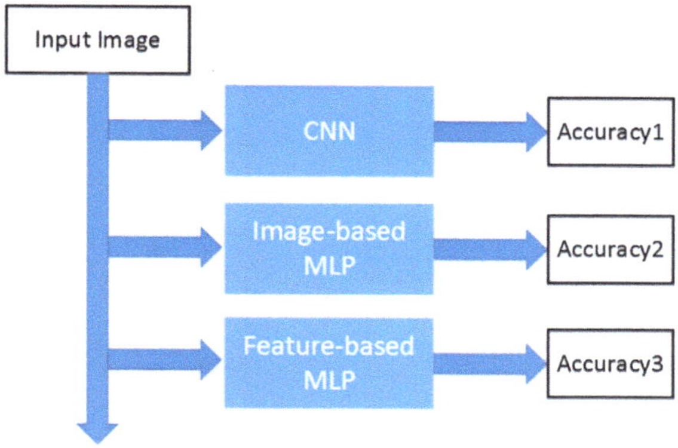

Fig. 4.3 Framework for comparing between automatic feature extraction in CNN and manual feature extraction using the same MLP classifier

neural network models for object classification, including CNN, image-based MLP (i.e. inputs of the MLP are raw image pixels), and feature-based MLP (i.e. inputs of the MLP are LBP features extracted from images).

(1) Automatic feature extraction using CNN. A modified version of the architecture of LeNet-5 CNN is used, which is composed of seven layers, including two convolutional layers, two pooling layers, two full connection layers, and one output layer. In the convolution layer, a set of learnable filters are used. Every filter is small spatially and covers the full depth of the input data. Given an image with a width W, height H and depth D, the learnable filters slide over the image spatially, resulting in a series of feature maps with a width and height of $W_1 = (W - F + 2P)/S + 1$ and $H_1 = (H - F + 2P)/S + 1$ respectively, where F refers to the spatial extent of neurons, P is the amount of zero padding, and S is the size of stride.

    The pooling layer operates independently on every depth slice of the input data and resizes it spatially using a max operation. For an image of $W \times H \times D$, the pooling layer reduces its size to $W_1 = (W - F)/S + 1$ for width and $H_1 = (H - F)/S + 1$ for height. After computing from all color channels, a max operation is performed. As a result, the feature matrix is reduced in the pooling layer. In the last layer, a MLP based fully connected network is used for object classification.
(2) Image-based MLP. The raw pixels of the full image are fed to a MLP classifier. The image data is firstly normalized and then a vector comprising the whole set of raw image pixels is fed to the MLP. The numbers of hidden neurons and training epochs are changed iteratively in the experiments.
(3) Feature-based MLP. A human engineering feature vector extracted from the image is fed to a MLP classifier. In this experiment, we use the LBP operator to generate the feature vector on the whole input image. In comparison to image-based MLP, feature-based MLP operates on a relatively small feature space due to a small number of features extracted from the image.

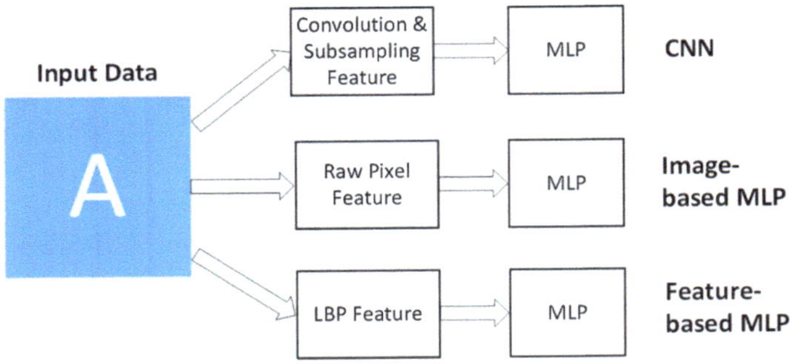

Fig. 4.4 Comparisons of feature extraction in CNN, image-based MLP and feature-based MLP

Figure 4.4 compares the architectures of CNN, image-based MLP, and feature-based MLP. It can be seen that the main differences of the three models are the features used, and the classifier is the same MLP.

4.3.3 Experimental Results

The experiments are based on three datasets, including the MNIST [11] dataset, a cow heat sensor dataset and the cropped roadside object dataset. Since the MNIST dataset has already been divided into 60,000 training and 10,000 test samples, we have used this data split. The cow dataset comprising around 50 heat sensor images was created to detect the change of body temperatures in cows. The image data is divided into two categories: changed and unchanged colors. Figure 4.5 displays two sample images showing two sensor devices with a color change and no color change respectively. The cropped roadside object dataset includes 650 regions manually cropped from the DTMR video frames and seven objects such as sky, tree, road, soil, and grass. For both cow heat senor and cropped roadside object datasets, we have used 75% data for training and the left 25% data for testing.

All the algorithms are developed on the Maltab platform. For image-based MLP and feature-based MLP, the default parameters are used and they are trained using a conjugate gradient descent based backpropagation algorithm.

Figure 4.6 shows the classification results obtained using CNN, image-based MLP and feature-based MLP on the three datasets with the same parameter settings. On the MNIST dataset, the three approaches have the highest accuracies of 99, 80, and 72%, respectively among using different number of epochs. The results suggest that inappropriate manual feature extraction will result in lower accuracy compared with automatic feature extraction. As for the cow dataset, the feature-based MLP is able to achieve 100% accuracy when 50 epochs are used for training, but its performance fluctuates wildly using different number of epochs. The CNN shows steadily increasing accuracy with the highest accuracy of 100%, while image-based MLP seems to have the lowest overall accuracy. It is noted that the number of

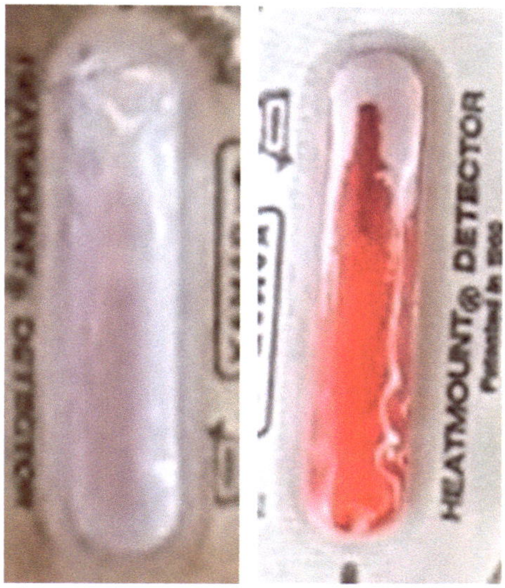

Fig. 4.5 Two images in the cow heat sensor dataset showing changed and unchanged colors

images in each class is relatively small (around 50) in the cow dataset. When evaluated on the cropped dataset, the three approaches perform similarly and the accuracy increases when a larger number of epochs is used until some points where a peak performance is obtained. The highest accuracy of 72.9% is achieved using feature-based MLP, followed by 72.7% using CNN and 70.6% using image-based MLP. CNN shows relatively lower classification accuracy compared with the results obtained on the MNIST and cow datasets, which is indicative of a higher level of difficulty of classifying objects in the cropped roadside dataset. It is interesting to see that, image-based MLP has got slightly higher accuracy than CNN. Although the performance of CNN may be rectified by changing its parameter settings, the result may be also true for recent extensions to CNN. The cropped roadside object dataset is a typical example showing the importance of appropriate feature extraction.

Although CNN has performed well, it is worth noting that CNN with automatic feature extraction is not the best classifier on the cropped roadside object dataset. The traditional image-based and feature-based MLPs have performed as well as or better than CNN in two out of three datasets. Thus, it is advisable to try both traditional MLP and CNN for object classification tasks, particularly when only small data is available.

4.3.4 Summary

This section has presented a comparison framework to investigate the impact of automatic feature extraction in CNN versus manual feature extraction on the performance of image classification tasks. Three models are used, including CNN with

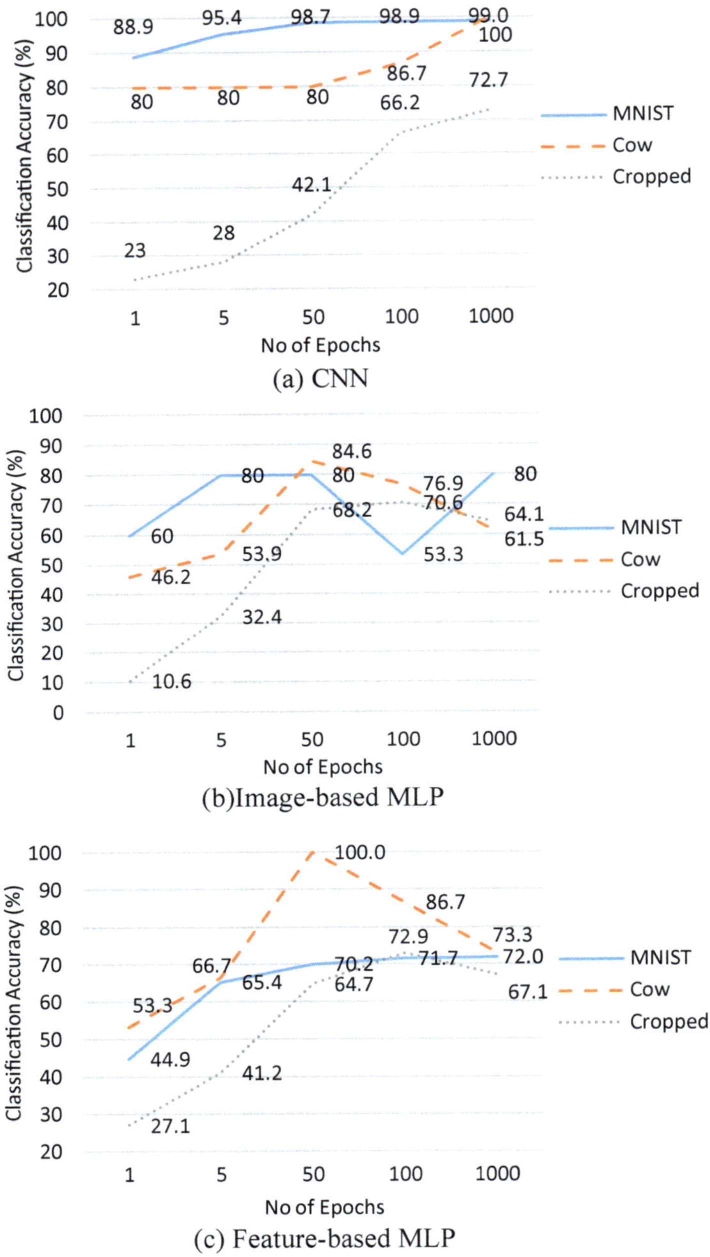

Fig. 4.6 Classification accuracy (%) obtained using CNN, image-based MLP and feature-based MLP. The MLP has 120 hidden neurons

automatic feature extraction, a traditional MLP with full raw image pixels, and a MLP with LBP feature extraction. The MLP is used as the classifier for all three models. Experiments with similar parameters settings have been conducted to systematically analyze the classification accuracies of the three models on three benchmark datasets. The results show that CNN with automatic feature extraction performs well but is not necessarily the best solution to small image classification tasks. For small datasets such as the cow and cropped roadside object datasets, a simple traditional MLP classifier with human-engineered features may serve equivalently or better than CNN.

4.4 Single Versus Ensemble Architectures

4.4.1 Introduction

The architecture of deep learning techniques is likely to have big impact on their performance. Recent work on deep learning favors going into 'deeper' learning architectures, but the increasingly complexity of 'deeper' architectures makes it difficult to train to optimize all parameters globally and use it for large scale image processing tasks. For instance, the deep Residual Network (ResNet) proposed by Microsoft Asia in 2015 [12] has a depth of up to 153 layers. One of the alternative to going deep is going wide and adopting an ensemble strategy, which can also be seen as increasing the depth by combining multiple deep learning networks into another layer. One example of the ensemble strategy is bagging [13], which provides insights into how an ensemble of classification and regression trees works when they were trained by taking random samples from the whole dataset.

In recent years, studies have shown that ensemble techniques can deliver outstanding performance in learning and reducing the test error. An ensemble model with five ConvNets [14] achieved a top-1 error rate of 38.1%, compared to the top-1 error rate of 40.7% given by the single model on the ImageNet 2012 classification benchmark. In [15], it was shown that by using an ensemble of six ConvNets, the top-1 error was reduced from 40.5 to 36.0%. A deep learning technique for robotic hand grasp detection was proposed in [16]. A two-step cascaded system comprising of two deep networks was used, in which the top detections from the first network were re-evaluated by the second network. Deep learning has been combined with an ensemble of neural networks [17], which was applied to a black box image classification problem with 130,000 unlabelled samples.

Although deep learning ensemble techniques have been applied to many application tasks, there lack of studies that systematically compare the performance of an ensemble of deep architectures with traditional classification techniques such as MLP. This section describes an ensemble of CNNs [18] and experimentally compares it with both an assemble of MLPs and a single CNN or MLP classifier to answer the following questions: (1) what is the performance of an ensemble of CNNs compared to an ensemble of traditional classifiers such as MLP? and (2) what is the performance of an ensemble of classifiers compared to a single classifier?

4.4.2 Comparison Framework

A systematic framework [18] is described in Fig. 4.7 to compare different architectures of CNN and MLP. Four models are considered: an ensemble of CNNs, an ensemble of MLPs, a single CNN, and a single MLP. The first two models are described below:

(1) An ensemble of CNNs. The ensemble architecture includes three individual CNNs as shown in Fig. 4.8. Each CNN in the ensemble contains standard layers such as convolutional layer, max pooling layer and fully connected layer. In the convolution layer, a set of filters is used and each filter has a variable size. A window size of 28×28 and a filter size of 5×5 are used in each CNN. The max-pooling layer operates independently at each depth slice of the input data and resizes it spatially using a max operator. Each CNN is trained separately and decisions of all CNNs are combined using majority voting. An appropriate number of CNNs in the ensemble is not further investigated because our purpose is not to find optimal ensemble parameters.
(2) An ensemble of MLPs. The raw pixels of the full image are taken as the input to an ensemble of three MLPs. The backpropagation algorithm is used for the training the MLP, and the numbers of hidden neurons and training epochs are changed iteratively. An overview of this method is presented in Fig. 4.8.

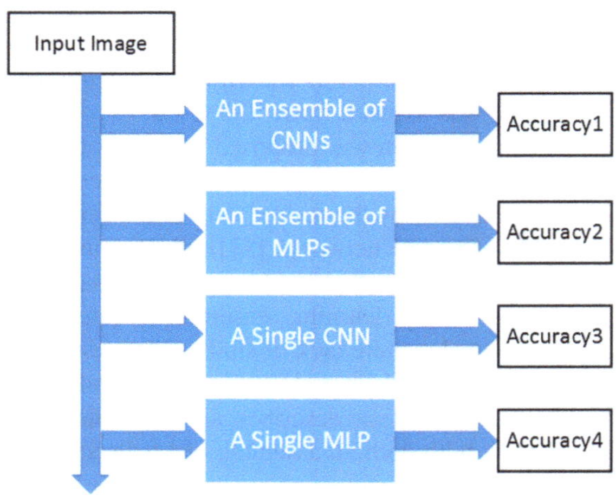

Fig. 4.7 Framework for comparing a single versus an ensemble of CNNs and MLPs

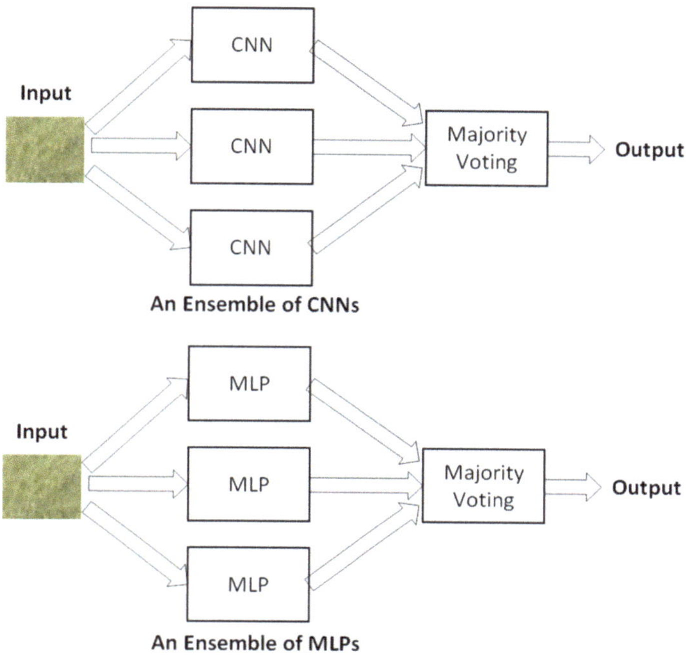

Fig. 4.8 Illustration of the architecture of an ensemble of CNNs (or MLPs)

4.4.3 Experimental Results

The experiments are conducted on three real-world datasets, including the MNIST, the cow heat dataset [19], and the cropped roadside object dataset. The training and test settings are the same as those in Sect. 4.3.3.

Tables 4.2 and 4.3 show the results obtained using an ensemble of CNNs and an ensemble of MLPs respectively on the three datasets. The results obtained by an ensemble of CNNs shows 99.3% accuracy which is higher than 95.2% accuracy obtained by an ensemble of image-based MLPs on the MNIST dataset. Similarly, an ensemble of CNNs shows higher accuracy (88.8% vs. 73.5%) than an ensemble of MLPs on the cropped roadside object dataset. While on the cow dataset, an ensemble of MLPs and an ensemble of CNNs have the same training and test

Table 4.2 Accuracy (%) using an ensemble of CNNs

Data	#Epochs			Train acc.	Test acc.
	CNN1	CNN2	CNN3		
MNIST	150	1000	1010	99.2	99.3
COW	1000	1050	1100	100	100
Cropped	1000	1050	1100	95.5	88.8

Table 4.3 Accuracy (%) using an ensemble of MLPs

Data	#Epochs			#Hidden neurons			Train acc.	Test acc.
	CNN1	CNN2	CNN3	CNN1	CNN2	CNN3		
MNIST	50	53	55	12	12	12	78.4	95.2
COW	100	100	101	6	16	6	100	100
Cropped	50	50	45	16	16	20	79.2	73.5

Table 4.4 Accuracy (%) using a single CNN

Data	Acc.	#Epochs		
		50	100	1000
MNIST	Train	92.0	94.3	98.2
	Test	98.7	98.9	99.0
Cow	Train	62.9	94.3	100
	Test	80.0	86.7	100
Cropped	Train	52.9	72.3	93.2
	Test	42.1	66.2	72.7

Table 4.5 Accuracy (%) using a single MLP

#Hidden neurons	MNIST			Cow			Cropped		
	#Epochs	Train	Test	#Epochs	Train	Test	#Epochs	Train	Test
6	1000	100	86.7	100	100	100	100	72.1	57.7
12	50	100	93.3	100	100	92.3	100	94.5	68.2
16	50	100	86.7	100	100	100	100	95.5	70.0
24	100	100	86.7	50	100	84.6	50	86.4	68.8
120	50	100	80.0	50	100	84.6	100	98.7	70.6

accuracies of 100%. The results indicate that it is more beneficial to build an ensemble of CNNs rather than an ensemble of MLPs.

Tables 4.4 and 4.5 show the results obtained using a single classifier of CNN and MLP respectively on the three datasets. For both CNN and MLP, an ensemble of classifiers is able to improve the test accuracy slightly compared to using a single classifier on the MNIST and cropped roadside object datasets. By contrast, a single CNN or MLP can achieve the same accuracy as an ensemble of CNNs or MLPs on the cow dataset, which is probably due to a small number of images and object categories in the cow dataset. For the MNIST dataset, an ensemble of CNN classifiers outperforms an ensemble of MLP classifiers, and a single CNN also outperforms a single MLP. It is noted that an ensemble of MLPs has lower accuracy than using a single CNN, probably because the parameters of the CNN have been optimized particularly for the MNIST dataset. On the cropped roadside object dataset, an ensemble of CNNs shows the highest test accuracy of 88.8%, which is followed by an ensemble of MLPs, while a single CNN and a single MLP perform the worst. The results indicate the benefit of considering an ensemble of CNN classifiers for more accurate results.

Although an ensemble of CNNs performs well in all experiments and produces the best results in terms of test accuracy (i.e. 99.3% on the MNIST dataset, 100% on the cow dataset and 88.8% on the cropped roadside dataset), it is worth noting that an ensemble of CNNs takes a longer time and more training epochs to achieve the highest accuracy. Traditional image based MLPs and an ensemble of MLPs perform as well as an ensemble of CNNs on only the cow dataset. An ensemble of CNNs has performed well for all datasets among the four models evaluated on the three datasets.

4.4.4 Summary

This section has presented an ensemble of CNNs and compared its performance with an ensemble of MLPs, a single CNN and a single MLP on three real-world datasets. For all four models, the raw pixels of the full image are used as inputs with similar training and test experimental conditions. Experimental results found that an ensemble of CNNs is the best performer, amongst the four models evaluated. Therefore, it is advisable to consider an ensemble of classifiers, particularly CNNs, for handling object classification problems in real-world datasets.

4.5 Deep Learning Network

4.5.1 Introduction

Recent efforts on video data analysis have largely shifted to the incorporation of contextual information to enforce constraints on the consistency of class labels to refine object classification results. Context is a statistical property of real world objects and contains critical information for assisting accurate class label inference in complex object classification tasks. Studies on human perception have found that human classification of objects has lower accuracy than machine classification when pixels outside objects are invisible [20]. There are two types of context—global context which collects image statistics from the holistic scene and local context which considers information from neighboring areas of interest. It is widely accepted that the class label of an image pixel should not only keep a high consistency with the local context conveyed by neighboring pixels, but also match the global context in terms of the whole scene semantics and layout.

The context is generally considered at two stages in existing studies on scene content analysis: feature extraction and label inference. Feature extraction collects a set of global or local contextual features to capture the intrinsic correlations between objects embedded in each type of scenes, which are often used in conjunction with visual features to improve class labelling accuracy. The commonly adopted contextual features include absolute location [21], relative location [22], directional spatial relationships [23, 24], and object co-occurrence statistics [25]. Hierarchical models, such as CNNs [26] and Contextual Hierarchical Models (CHMs) [27], have shown promising results in learning visual and contextual feature representations from raw image pixels. Label inference aims to impose contextual consistency of predicted class labels in an energy minimization process using graphical models, such as CRFs [26, 28, 29], MRFs [30], and region boundary based energy functions [16].

In recent years, deep learning techniques have shown advantages in extracting robust contextual features from raw image pixels. The widely used CNNs utilize convolutional and pooling layers to progressively extract abstract and contextual patterns, which are jointly modelled with the MRF or CRF inference to generate strong predictors. Farabet et al. [26] applied hierarchical CNN features into CRFs for label inference. Schwing and Urtasun [31] passed the error of the CRF inference backward to CNNs. However, the CRF inference is completely independent from CNN training. To solve this problem, Zheng et al. [32] formulated the CRF inference as recurrent neural networks and integrated them in a unified framework. Although CNNs achieve promising results, they have two drawbacks: (1) they often get confused by visually similar pixels due to the consideration of limited context; and (2) they are unable to automatically adapt to the image content due to the dependency of per-learnt system parameters. To encode higher order context, the recurrent CNN [33] feeds back the outputs of CNNs to the input of another instance of the same network, but it works only on sequential data. The recursive context propagation network [34, 35] recursively aggregates local contextual information up to the entire image and then disseminates the aggregated information back to local features, but it is still impacted by impurity labels of superpixels and may suffer from severe error propagation from one level to another. The CHMs [27] incorporate multi-resolution contextual information by integrating the outputs of a series of classifiers in a hierarchy, but the contextual information is based on downsampled images where a significant proportion of information is lost.

The approaches introduced above have produced promising results for pixel labelling in natural scenes. However, they generally have three drawbacks: (1) the contextual features either completely discard absolute spatial coordinates of objects, which also carry important contextual information for scene parsing, or excessively preserve all absolute coordinates, which requires a large amount of training data to ensure reliable performance. (2) The MRF and CRF label inference models have limited capacity of capturing global context such as long-range label dependencies between objects in the entire image. They primarily focus on the consistency of labels in local neighborhoods and thus have limited power of ensuring global coherence. (3) Inability to adapt to the local property in new scenes, which is crucially important to overcome variability of objects and changes in the environment [36].

In this section, we describe a deep learning network [37], which is able to learn both short and long term contextual features to improve object segmentation accuracy on real-world data. Rather than performing in a hierarchical manner, the deep learning network borrows the concept of multiple layers in deep learning to obtain robust contextual features in image blocks and integrate them with visual features for robustly labelling image pixels to different objects.

4.5.2 Deep Learning Network

4.5.2.1 Problem Formulation

Let $I(v) \in R^3$ be an image comprising a set of pixels v, the purpose of object segmentation is to assign every pixel v into one of class labels $C = \{c_i | i = 1, 2, \ldots, M\}$ and M is the number of all classes. For superpixel based object classification, let $S(v) = \{s_j | j = 1, 2, \ldots, N\}$ indicate the set of superpixels oversegmented from I and N be the number of all superpixels, their corresponding visual are features are $F^v = \{f_j^v | j = 1, 2, \ldots N\}$, local contextual features are $F^l = \{f_j^l | j = 1, 2, \ldots N\}$ and global contextual features are $F^g = \{f_j^g | j = 1, 2, \ldots N\}$. The task is then transferred to labelling all pixels v in s_j into a class $c_i \in C$ and $v \in s_j$, and the conditional probability of assigning a correct label c_i to s_j can be expressed as:

$$\begin{aligned} &P(c_i | s_j, W) = P\left(c_i \middle| f_j^v, f_j^l, f_j^g; W^v, W^l, W^g\right) \\ &\text{s.t.} \sum_{1<i<M} P(c_i | s_j) = 1 \end{aligned} \tag{4.1}$$

where, $W = \{W^v, W^l, W^g\}$ indicate weight parameters for F^v, F^l and F^g features respectively, which can be learnt from the training data.

The final target is to obtain a model which can maximize the sum of conditional probabilities of assigning correct labels for all superpixels:

$$P(C|S, W) = \max_{s_j \in S \,\&\, c_i \in C} P(c_i | s_j, W) \tag{4.2}$$

Now, there are two tasks: (1) how to obtain the local and global contextual features that incorporate both short- and long-range label dependencies between objects, and are adaptive to local properties of test images? and (2) how to seamlessly integrate visual and contextual features to obtain the maximized conditional probabilities for all superpixels.

4.5.2.2 Network Architecture

As shown in Fig. 4.9, the Deep Learning (DP) network architecture [37] for solving the two tasks in the problem formulation is composed of three layers: (1) the visual feature prediction layer builds class-semantic supervised classifiers to predict class probabilities of all superpixels based on their visual features. (2) The contextual voting layer obtains local and global Context Adaptive Voting (CAV) features of each superpixel based on its most probable class and the corresponding Object Co-occurance Priors (OCPs). (3) The integration layer jointly models the correlations of visual and CAV features to derive a final class label for each superpixel.

Specifically, the network takes superpixel-level visual features as inputs and outputs a class label for each superpixel. The second layer accepts class probabilities predicted in the first layer and integrates the most probable class with OCPs (learnt from the training data) to learn image-dependent CAV features, which reflect class label preferences for each superpixel voted by global or local context in a test image:

$$V^l(C|s_j) = \psi^l(P^v(C|s_j), OCP) \tag{4.3}$$

$$V^g(C|s_j) = \psi^g(P^v(C|s_j), OCP) \tag{4.4}$$

where ψ^l and ψ^g stand for the voting functions for global and local context respectively. The CAV features adaptively capture short- and long-range label dependencies of superpixels, as well as the local properties of the test image.

To integrate contextual and visual features, we consider visual features, local and global CAV features as three individual components and normalize CAV features into class probabilities. We then jointly model the correlations of class probabilities of visual and CAV features in the third layer using a set of optimized weights of neurons:

$$P(C|s_j) = \mathcal{H}\left(\overbrace{P^v(C|s_j)}^{visual feature}, \overbrace{P^l(C|s_j)}^{local\,CAV}, \overbrace{P^g(C|s_j)}^{global\,CAV}\right) \tag{4.5}$$

$$P^T(C|s_j) = \cup_{1<i<M} P\left(c_i \middle| f_j^T; W^T\right) \tag{4.6}$$

where $\mathcal{H}$ stands for the function for jointly modelling three types of class probabilities $P^T(C|s_j)$ of s_j based on features f_j^T and the corresponding weights W^T, $T \in \{v, l, g\}$.

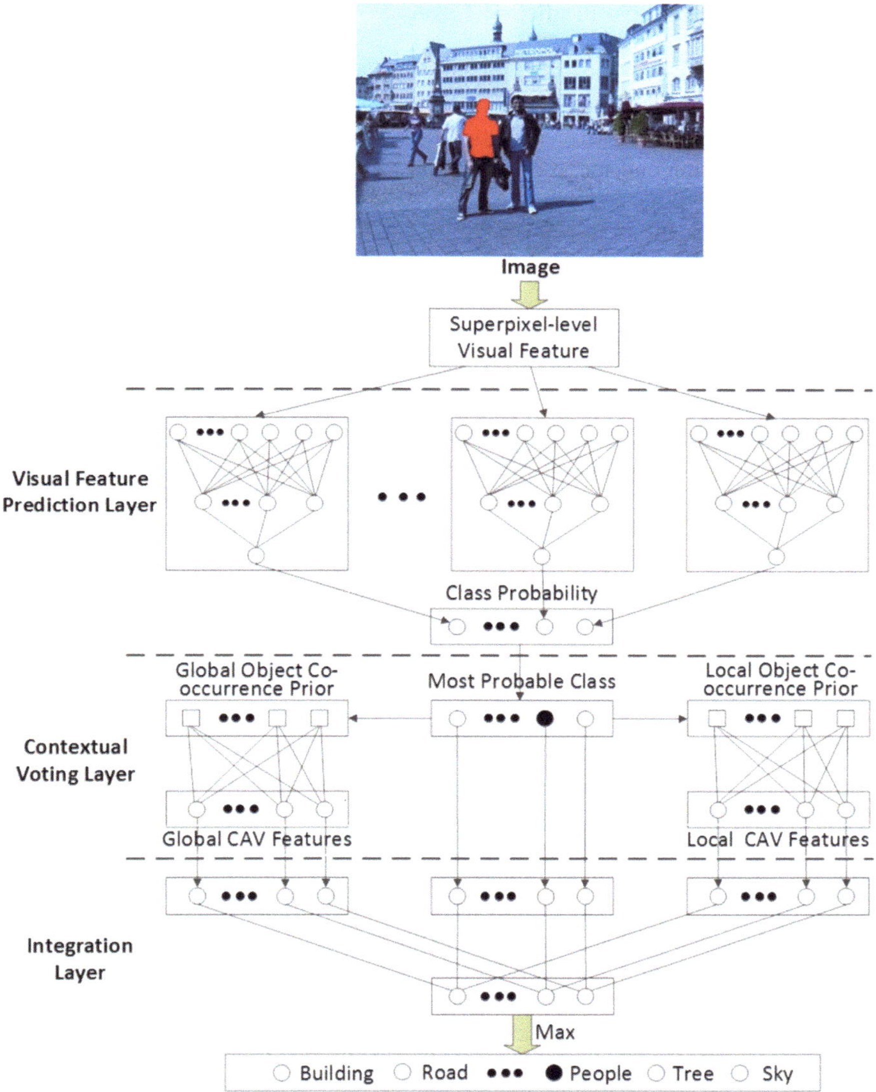

Fig. 4.9 Framework of the three-layer deep learning network architecture. The prediction layer takes superpixel-level visual features as inputs and predicts class probabilities of each superpixel using class-semantic classifiers. The context voting layer calculates global and local Context Adaptive Voting (CAV) features by casting votes from superpixels in spatial blocks and from adjacent superpixels respectively based on the most probable class and Object Co-occurance Priors (OCPs). The OCPs (shown in *rectangles*) are collected for all block pairs from the training data. The integration layer integrates visual feature based class probabilities, global and local CAV features to produce a class probability vector. Each superpixel is finally assigned to a class label using a majority voting strategy

4.5.2.3 Visual Feature Prediction Layer

For an input image, the visual feature prediction layer incorporates multiple class-specific classifiers to obtain an approximate prediction of the probabilities of all superpixels belonging to each class based on a set of visual features. It serves as an initial prediction and forms a foundation for generating CAV features based on the context in the image.

For the *j*th superpixel s_j, its class probability for the *i*th class c_i can be obtained:

$$P^v(c_i|s_j) = \o_i\left(f_{i,j}^v\right) = fn(w_{1,i}f_{i,j}^v + b_{1,i}) \tag{4.7}$$

where $f_{i,j}^v$ is the visual features of s_j extracted for the *i*th class c_i, $\o_i$ is the trained binary classifier for c_i, fn indicates the prediction function of $\o_i$, and $w_{1,i}$ and $b_{1,i}$ are trainable weights and constant parameters respectively.

For all M classes, a class probability vector for s_j can be obtained:

$$P^v(C|s_j) = [P^v(c_1|s_j), \ldots, P^v(c_i|s_j), \ldots, P^v(c_M|s_j)] \tag{4.8}$$

The above vector includes the likelihoods of each superpixel s_j belonging to all classes C. We can assign s_j to the class which has the maximum probability:

$$s_j \in \hat{c} \quad if\ P^v(\hat{c}|s_j) = \max_{1<i<M}(P^v(c_i|s_j)) \tag{4.9}$$

Rather than training a single multi-class classifier for classifying all classes in the same model, a series of class-specific one-versus-all classifiers are designed and trained for each class. The use of class-specific classifiers has three advantages: (1) allowing the selection of most discriminative class-specific features specifically for each class, (2) focusing on training a more powerful and effective classifier for a specific class at each time, and (3) handling the problem of unbalanced training data between classes particularly in natural datasets where many rarely occurring but important classes often exist. For real-world datasets, where the distribution of pixels may be heavy-tailed to several common classes, training a multi-class classifier has the risk of completely ignoring rare classes and being favourably biased towards those common classes.

4.5.2.4 Contextual Voting Layer

The contextual voting layer is the central part of the deep learning network, and its aim is to learn the CAV features, which capture long- and short-range label dependencies between objects and are adaptive to the local property in test images. There are two main processing steps:

(1) Calculation of Object Co-occurrence Priors (OCPs) from the training data.

The OCPs encode prior spatial correlations between objects regarding their class label distributions in spatial blocks of training images. To effectively capture the prior context knowledge about a specific type of scene, the OCPs take into account four types of inter-class spatial correlations, including object co-occurrence frequency, relative location, absolute location, and directional spatial relationships. These correlations carry important semantic information about the scene and can be used to enforce contextual constraints on class labels predicted using visual feature based classifiers. Figure 4.10 exemplifies the differences between OCPs and relative/absolute location in modelling of inter-class spatial correlations.

The calculation of OCPs is described as follows. To consider both relative and absolute locations of objects, the image I is firstly partitioned into a set of equally distributed blocks $B = \{B_k | k = 1, 2, \ldots, K\}$, where K is the number of blocks. The use of blocks is aimed at keeping a good trade-off between relative and absolute location contextual information in which the relative location offsets of objects are encoded by spatial relationships between blocks, while the absolute location is

Fig. 4.10 Illustration of the differences between relative location prior, absolute location prior, and OCP in modelling of three cases of a spatial correlation "left sky and right building" (best viewed in color)

preserved by spatial coordinates of each block. The spatial distributions of all blocks also preserve directional spatial relationships between blocks, e.g. left and right spatial relationships. As objects may present at any location in a scene, both global and local OCPs are created to represent long- and short-range label dependencies of two objects co-occurring in the scene, and they contain prior knowledge about global and local context respectively embedded in training scenes of a specific type of object.

(a) Global OCP. Assume a pixel with a class label $\hat{c}$ occur in a block B_{k2}, a matrix $\mathcal{M}_{c|\hat{c}}(k_1, k_2)$, $k_1 \neq k_2$ stands for the probability that a pixel with a class c occurs in a block B_{k1} and $\mathcal{M} \in \mathcal{R}^{M \times M \times K \times (K-1)}$. The matrix is normalized to ensure a conditional probability distribution over all classes in each block, i.e. $\sum_{c=1}^{M} \mathcal{M}_{c|\hat{c}}(k_1, k_2) = 1$. For every pair of blocks, a matrix with $M \times M$ elements can be generated and each element indicates the pixel co-occurrence frequency between two objects that are located in the two different blocks. Figure 4.11 visually illustrates the calculation of an object co-occurrence matrix for two blocks. For a superpixel in a block B_{k1}, the global OCP reflects the confidence in its class label with the support of contextual information from all superpixels in other $K-1$ blocks $\cup(B_{k2})k_2 \neq k_1$.

(b) Local OCP. Given an adjacent superpixel s_p (of a superpixel s_j) with a class label $\hat{c}$, a matrix $\overline{\mathcal{M}}_{c|\hat{c}}(s_j, s_p)$ represents the probability that s_j with a class c will occur and $\overline{\mathcal{M}}_{c|\hat{c}}(s_j, s_p) \in \mathcal{R}^{M \times M}$. The local matrix is calculated for every pair of adjacent superpixels on the training data, and all superpixel pairs may appear at any location of the scene because the spatial block is not considered during the calculation. It compensates for the "self-support" contextual information for superpixels within the same block that is not considered in the global OCP. For

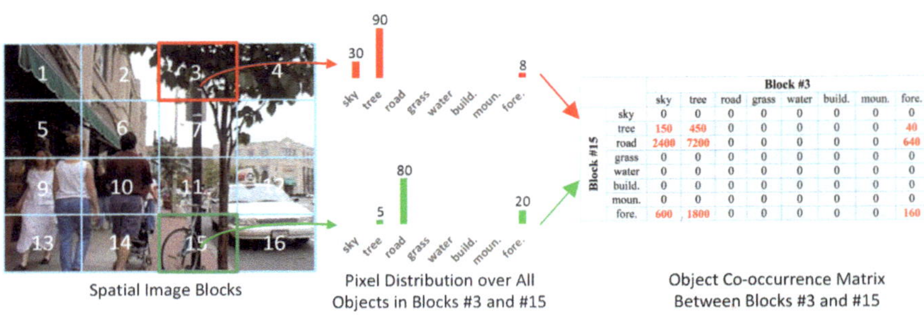

Fig. 4.11 Illustration of the calculation of the global OCP in an image from the Stanford background dataset. The pixel distributions over all classes are calculated for each of all spatial image blocks, and an object co-occurrence matrix is then formed for every two-block pair (#3 and #15 in the example) to reflect the long-range label dependencies between objects within two blocks

a superpixel, local OCP accounts for the confidence in its class label with the support of contextual information from its neighboring superpixels $\bar{S}_j = \cup (s_p)$.

For both global and local OCPs, the occurrence frequency of each class is calculated by counting pixels within all superpixels of this class. The class label of each superpixel is obtained based on the majority vote of the ground truth pixel labels. Because of variations in the shape and size of segmented superpixels, a superpixel may present across the boundaries of multiple spatial blocks. Thus each superpixel s_j in an image I is assigned to a unique block B_k:

$$s_j \in B_k \quad \text{if}\left(\bar{x}_j, \bar{y}_j\right) \in B_k \tag{4.10}$$

where $\left(\bar{x}_j, \bar{y}_j\right)$ is the centroid of all pixels within s_j.

(2) Generation of CAV features in a testing image.

Once global and local OCPs are obtained, we can proceed to describe how to incorporate them into the calculation of CAV features for all superpixels in a test image. For each of all superpixels, we first obtain the most probable class using (4.9) and the corresponding spatial block using (4.10), and then adopt a voting strategy to obtain CAV features for the superpixel based on global and local OCPs. Specifically, each superpixel in a spatial block casts a vote to the class probabilities of superpixels in other blocks based on global OCP $\mathcal{M}_{c|\hat{c}}(k_1, k_2)$, as well as a vote to the class probabilities of its neighboring superpixels based on local OCP $\overline{\mathcal{M}}_{c|\hat{c}}(s_j, s_p)$.

Given a superpixel s_j in a block B_{k1}, let $S = \{s_q | q = 1, 2, \ldots, Q\}$ be the set of superpixels in all the rest blocks $\cup (B_{k2}) k_1 \neq k_2$ and s_q be the qth element of S, the s_j receives Q votes from all $s_q \in S$:

$$V^g(C|s_j) = \sum_{1 < k_2 < K; k_1 \neq k_2} \sum_{s_q \in B_{k2}; s_q \in \hat{c}} w_{2,q} \times \mathcal{M}_{c|\hat{c}}(k_1, k_2) \tag{4.11}$$

where $\hat{c}$ is the most probable class of s_q, $P(\hat{c}|s_j)$ is the corresponding class probability, $w_{2,q} = P(\hat{c}|s_j) \times \mathbb{C}(s_q)$ is a weight given to the vote from s_q and $\mathbb{C}(s_q)$ is the total number of pixels in s_q.

Since s_q can have M possible class labels, the resulting global CAV features $V^g(C|s_j)$ can be written as:

$$V^g(C|s_j) = [V^g(c_1|s_j), \ldots, V^g(c_i|s_j), \ldots, V^g(c_M|s_j)] \tag{4.12}$$

where $V^g(c_i|s_j)$ indicates contextual label votes of being the c_i class for s_j based on global OCP.

Let $\bar{S}_j = \{s_p|p = 1, 2, \ldots, P\}$ be the set of adjacent superpixels of s_j and s_p be the pth member of $\bar{S}_j$, the s_j receives P votes from its neighbors s_p:

$$V^l(C|s_j) = \sum_{1<p<P; S_p \in \hat{c}} w_{2,p} \times \overline{\mathcal{M}}_{c|\hat{c}}(s_j, s_p) \tag{4.13}$$

where $\hat{c}$ is the most probable class of s_p and the weight $w_{2,p}$ is calculated in the same way as $w_{2,q}$ in (4.11). Using a way similar to (4.12), local CAV features indicating contextual label votes of all classes for s_j can be obtained based on local OCP:

$$V^l(C|s_j) = [V^l(c_1|s_j), \ldots, V^l(c_i|s_j), \ldots, V^l(c_M|s_j)] \tag{4.14}$$

Now, we have calculated two types of CAV features for each superpixel based on global and local OCPs respectively, and they represent the confidence in class labels for a superpixel using global and local context respectively in the scene.

4.5.2.5 Integration Layer

The integration layer aims to obtain context-sensitive classification of all superpixels in a testing image by learning an optimized set of neuron weights to seamlessly integrate visual feature based class probabilities $P^v(C|s_j)$, local CAV features $V^l(C|s_j)$ and global CAV features $V^g(C|s_j)$. For this aim, we learn class-specific optimized weights of the corresponding neurons to best describe the correlations between each class and the three predictive terms. These weights inherently account for different contributions of visual features and contextual cues in predicting the class labels of superpixels in the testing image. The weight optimization process is conducted in a similar way to multiple linear regression models.

To keep consistency with class probabilities $P^v(C|s_j)$, both global and local CAV features are normalized into probabilities whose values over all classes sum to 1:

$$P^g(c_i|s_j) = V^g(c_i|s_j) / \sum_{1<i<M} V^g(c_i|s_j) \tag{4.15}$$

$$P^l(c_i|s_j) = V^l(c_i|s_j) / \sum_{1<i<M} V^l(c_i|s_j) \tag{4.16}$$

For each of all classes, a connection of its corresponding neurons from P^v, P^l and P^g is then established to integrate three probabilities for the prediction of the probability of the ith class c_i for superpixel s_j:

$$P(c_i|s_j) = b^c_{3,i} + w^v_{3,i} \times P^v(c_i|s_j) + w^l_{3,i} \times P^l(c_i|s_j) + w^g_{3,i} \times P^g(c_i|s_j) \tag{4.17}$$

where $b^c_{3,i}$ is a constant value for c_i, and $w^a_{3,i}$, $w^l_{3,i}$ and $w^g_{3,i}$ are weights for $P^a(c_i|s_j)$, $P^l(c_i|s_j)$ and $P^g(c_i|s_j)$ respectively in the third layer. The weights are learnt to minimize the sum of squared deviations of likelihood scores over our labeled training data:

$$\min\left(\sum_{j=1}^{J} \left(P(c_i|s_j) - P'(c_i|s_j)\right)^2\right) \tag{4.18}$$

where J is the total number of all superpixels in the training data. $P'(c_i|s_j) \in \{1, 0\}$ indicates a ground truth class label of c_i or non-c_i respectively. The predicted probabilities are assumed to have a normal distribution.

For all M classes, a series of connected neuron models with different weights are learnt:

$$P(C|s_j) = [P(c_1|s_j), \ldots, P(c_i|s_j), \ldots, P(c_M|s_j)] \tag{4.19}$$

Finally, the superpixel s_j is labelled as the class $\hat{c}$ that has the highest probability across all classes using a majority voting strategy:

$$s_j \in \hat{c} \text{ if} P(\hat{c}|s_j) = \max_{1<i<M} P(c_i|s_j) \tag{4.20}$$

The training and test procedures of the deep learning network are summarized in Algorithm 4.1.

Algorithm 4.1: Training and test proceduresinthe deep learning network.

Input: image I; superpixels $S = \{s_j | j = 1,2,\ldots,N\}$; class labels $C = \{c_i | i = 1,2,\ldots,M\}$; spatial blocks $B = \{B_k | k = 1,2,\ldots,K\}$.
Output: class label for s_j

// Initial prediction using visual features
Foreach superpixels $s_j \in S$
  Foreach classes $c_i \in C$
    Extract visual features $f_{i,j}^v$;
    Get class probability: $P^v(c_i|s_j) = fn(w_{1,i} f_{i,j}^v + b_{1,i})$
  End
  Get the most probable class: $\hat{c} = \max_{1<i<M}(P^v(c_i|s_j))$
End
// Assign each superpixel to a spatial block
Foreach superpixels $s_j \in S$
  Get the centroid of s_j: $(\bar{x}_j, \bar{y}_j)$
  Assign s to a block: $s_j \in B_k$ if $(\bar{x}_j, \bar{y}_j) \in B_k$
End
// Calculate OCP
Foreach blocks $B_{k2} \in B$
  Foreach blocks $B_{k1} \in B$ & $k_1 \neq k_2$
    Get global OCP for classes c and $\hat{c}$: $\mathcal{M}_{c|\hat{c}}(k_1, k_2)$
  End
End
Foreach superpixels $s_j \in S$
  Foreach neighbours of s_j: $s_p \in \bar{S}_j$
    Get local OCP for classes c and $\hat{c}$: $\bar{\mathcal{M}}_{c|\hat{c}}(s_j, s_p)$
  End
End
// Calculate VOC features
Foreach superpixels $s_j \in S$

Foreach classes $c_i \in C$

Calculate global CAV features:

$$V^g(C|s_j) = \sum_{1<k_2<K;k_1\neq k_2} \sum_{s_q \in B_{k2};\, s_q \in \hat{c}} w_{2,q} \, \mathrm{X} \mathcal{M}_{c|\hat{c}} \, (k_1, k_2)$$

Calculate local CAV features:

$$V^l(C|s_j) = \sum_{1<p<P;\, s_p \in \hat{c}} w_{2,p} \, \mathrm{X} \, \bar{\mathcal{M}}_{c|\hat{c}}(s_j, s_p)$$

End

End

// Final prediction integrating visual and VOC features

Foreach superpixels $s_j \in S$

Foreach classes $c_i \in C$

Convert VOC to probability:

$$P^g(c_i|s_j) = V^g(c_i|s_j) / \sum_{1<i<M} V^g(c_i|s_j)$$

$$P^l(c_i|s_j) = V^l(c_i|s_j) / \sum_{1<i<M} V^l(c_i|s_j)$$

Integrate visual and CAV features:

$$P(c_i|s_j) = b^c_{3,i} + w^v_{3,i} \, \mathrm{X} P^v(c_i|s_j) + w^l_{3,i} \, \mathrm{X} P^l(c_i|s_j) + w^g_{3,i} \, \mathrm{X} \, P^g(c_i|s_j)$$

End

Assign s_j to a class with the maximum probability:

$$s_j \in \hat{c} \text{ if } P(\hat{c}\,|s_j) = \max_{1<i<M} P(c_i|s_j)$$

End

4.5.3 Experimental Results

The performance of the deep learning network using the CAV features is evaluated on three widely used datasets for scene parsing: Stanford background, MSRC and SIFT Flow. We also compare our results with the accuracies reported in the literature by the state-of-the-art scene parsing algorithms.

4.5.3.1 Experimental Set-Ups

(1) *Superpixel-level visual features*. The features consist of a set of color, geometric and texture features [38], including the mean and standard deviations of RGB colors over the pixels of each superpixel (2×3 dimensions), the top height of superpixel bounding box to the image height (1 dimension), the mask of superpixel shape over the image (8×8 dimensions), 11-bin histograms of RGB colors (11×3 dimensions), 100-bin histogram of textons (100 dimensions), and 100-bin histogram of dense SIFT descriptors (100 dimensions) over the superpixel region. In addition, we also obtain RGB, texton and SIFT histograms (233 dimensions) over the superpixel region dilated by 10 pixels. The SIFT descriptors are computed using 8 orientations and 4 scales filters, and the textons are defined as the clustered 8-dimensional responses to rotationally invariant Maximum Response 8 filter bank using the K-means clustering algorithm. The final visual feature vector is composed of 537 elements. We further obtain class-specific subsets of top 50 features for each class using the minimum redundancy maximum relevance algorithm [39].

(2) *System parameters*. Superpixels are obtained using a graph-based algorithm [40], and its parameters are set based on [38], i.e. $\sigma = 0.8, min = 100$, $k = 200 \times \max(1, sqr(D_I/640))$, and D_I is the larger dimension (height or width) of an image I. We evaluate two commonly used ANN and SVR classifiers for obtaining visual feature based class probability. The ANN has three layers of 50-16-1 neurons for each class, and the SVR uses a RBF kernel. The number of spatial blocks is set to 36, i.e. 6 in width and 6 in height.

(3) *Evaluation metrics*. We use three evaluation metrics: *global accuracy* which is the ratio of correctly classified pixels to the total test pixels, *average accuracy* which is the mean of category-wise pixel accuracy, and *class accuracy* which is the pixel accuracy for each category. The average accuracy treats the classification of each class equally regardless of their occurring frequency. The class accuracy indicates the performance for each object.

(4) *Evaluation strategies*. For the Stanford background dataset, we follow the evaluation procedure in [41], namely fivefold cross validations are conducted to obtain classification accuracy: 572 images are randomly selected for training and 143 images for testing in each fold. Similar to the Stanford background dataset, fivefold cross validations are used for obtaining the classification accuracy on the MSRC 21-class dataset. For the SIFT Flow dataset, the same train/test data split as [42] is used: 2488 training and 200 test images.

4.5.3.2 Performance on the Stanford Background Dataset

Table 4.6 shows global and average accuracies of the deep learning network compared with accuracies achieved by state-of-the-art methods. The network achieves global accuracies of 80.6 and 81.2% using ANN and SVR classifiers respectively, and the performance is competitive to the state of the art. A lower class accuracy of 72% is observed for the deep learning network compared with 76 and 79% accuracies in existing methods [35, 43], indicating that the deep learning network tends to focus more on common classes with a large proportion of training pixels, which is within our expectation as the CAV features were generated based on class pixel distributions in the training data. It is believed that the more training pixels for a specific class, the more reliable its OCPs and the higher accuracy of the corresponding CAV features. The network significantly outperforms visual feature based ANN and SVR classifiers with increases of more than 10 and 38% accuracies respectively, confirming the advantage of the CAV features in reducing misclassification and refining classification labels. Figure 4.12 visually shows qualitative results.

Table 4.7 shows the confusion matrix for eight classes. We can observe that sky, building and road are three easiest objects for correct classification with more than 87% accuracies, whereas mountain and water are the most difficult ones with less than 56% accuracies. The results are consistent with those in existing studies [41, 44, 45], where sky and mountain have the highest accuracies of around 94% and lowest accuracies of around 14%. For mountain, 24.6% of all pixels are misclassified to tree, probably due to the overlap in color features. Similarly, a large proportion (20 and 17.3%) of water pixels is misclassified to road and foreground objects. Among all classes, the foreground objects lead to the most overall misclassification of other classes, which is largely due to big variations in the appearance of foreground objects in natural scenes.

4.5.3.3 Performance on the MSRC Dataset

Table 4.8 shows the performance of the deep learning network compared to state-of-the-art methods. Our network using ANN and SVR classifiers achieves global accuracies of 82.1 and 85.5% respectively on the MSRC dataset. The performance

Table 4.6 Performance (%) comparisons with previous approaches on the Stanford background dataset

Refs.	Global Acc.	Avg. Acc.
Gould et al. [41]	76.4	–
Sharma et al. [34]	81.8	73.9
Shuai et al. [51]	81.2	71.3
Sharma et al. [35]	82.3	79.1
Visual feature (ANN)	69.7	55.8
DP network (ANN)	80.6	72.1
Visual feature (SVR)	43.1	35.2
DP network (SVR)	81.2	71.8

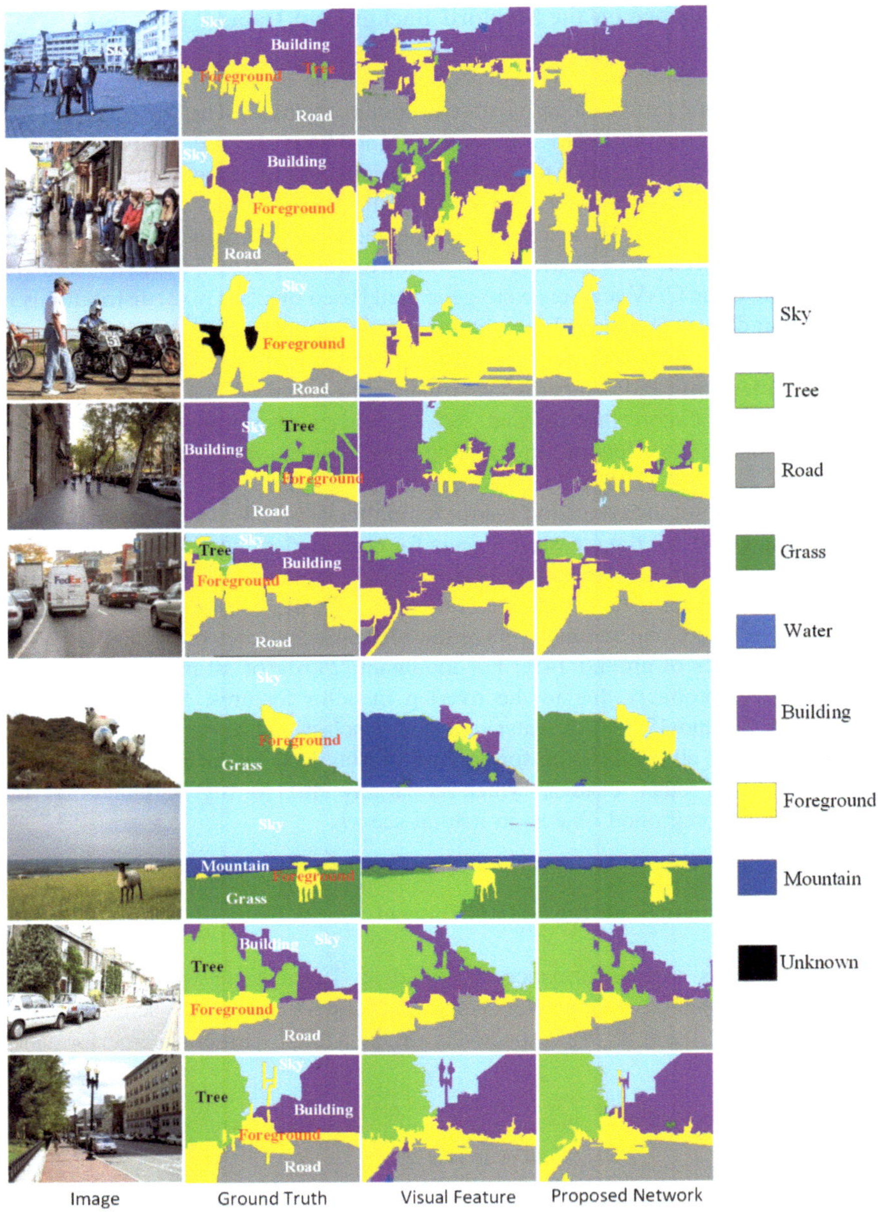

Fig. 4.12 Qualitative results on the Stanford background dataset (best viewed in color). Compared with using visual features alone, the deep learning network shows more robust scene parsing results by utilizing CAV features, with many misclassifications successfully removed

Table 4.7 Confusion matrix for eight objects on the Stanford background dataset (SVR, global accuracy 81.2%)

	Sky	Tree	Road	Grass	Water	Building	Mountain	Foreground
Sky	**91.2**	3.9	0.1	0.1	0.1	3.3	0.1	1.2
Tree	2.7	**74.3**	1.1	1.2	0.2	14.4	0.2	6.0
Road	0.1	0.6	**87.2**	1.8	0.3	2.8	0	7.1
Grass	0.3	5.2	10.6	**64.4**	1.0	1.8	0.8	15.9
Water	3.2	0.6	20.0	1.7	**55.6**	1.0	0.7	17.3
Building	1.4	4.5	1.3	0.2	0.3	**88.6**	0.1	3.7
Mountain	7.6	24.6	4.7	4.7	2.0	8.7	**37.2**	10.5
Foreground	1.3	4.2	6.1	1.1	0.7	11.1	0.1	**75.4**

The bold figure indicates class accuracy for each object.

using SVR outperforms all benchmark approaches except for [45] which reported 87% global accuracy. Among all methods, the deep learning network obtains the highest average accuracies of 82 and 85% over all classes, indicating an overall high and balanced performance between all classes. It also keeps the highest class accuracies for 11 out of all 21 classes and this confirms the benefit of the deep learning network for obtaining high accuracies for most classes. Although higher accuracies are obtained using ANN for most classes, a higher global accuracy is obtained using SVR, which is probably due to unbalanced pixel distributions across classes. Figure 4.13 displays classification results on sample images, which confirm a better performance of using CAV features than using visual features alone.

Table 4.9 presents a full confusion matrix summarizing pixel-wise class accuracies over all 21 classes. As we can see, book, flower, building, car and tree are the five easiest objects for correct recognition with more than 91% accuracies, whereas road, water, bird and boat are the most difficult ones with less than 77% accuracies. The 'object' classes such as chair, flower, cat, dog, sign, book and sheep lead to little confusion to other classes, while the 'background' classes such as building, grass, tree, sky and road result in most misclassification to other classes. More than 10% pixels of road, sky and face are misclassified to grass, building and body respectively, probably due to their neighbor relationships in most natural scenes and thus local CAV features tend to impose strong constraints on the spatial consistency of their labels.

4.5.3.4 Performance on the SIFT Flow Dataset

Table 4.10 compares the accuracy of the DP network with state-of-the-art methods. The network shows higher performance than other methods compared and it significantly improves the state-of-the-art global accuracy from 80.9 to 87.0% using either an ANN or SVR classifier. The significant improvement may due to a much larger number of training images on the SIFT Flow dataset compared to the Stanford background and MSRC datasets, which is crucially important for collecting reliable OCPs and calculating accurate CAV features. In addition, the DP

Table 4.8 Performance (%) comparisons with previous approaches on the MSRC dataset

Refs.	Global	Avg.	Building	Grass	Tree	Cow	Sheep	Sky	Aeroplane	Water	Face
[52]	72	58	62	98	86	58	50	83	60	53	74
[53]	81	74	67	96	88	82	83	91	81	66	89
[54]	–	62	74	93	84	61	60	79	55	75	75
[55]	84	79	67	89	85	93	79	93	84	75	79
[56]	84	81	67	95	92	91	90	95	96	73	88
[45]	87	78	81	96	89	74	84	99	84	92	90
DP ANN	82	82	94	82	92	89	87	56	89	68	84
DP SVR	85	85	94	87	91	89	89	86	83	71	82

Refs.	Car	Bicycle	Flower	Sign	Bird	Book	Chair	Road	Cat	Dog	Body	Boat
[52]	63	75	63	35	19	92	15	86	54	19	62	07
[53]	79	92	79	70	45	93	80	78	78	41	72	13
[54]	62	75	81	71	36	72	25	75	52	39	49	10
[55]	87	89	92	71	46	96	79	86	76	64	77	50
[56]	76	94	90	76	57	84	69	82	89	60	84	44
[45]	86	92	98	91	35	95	53	90	62	77	70	12
DP ANN	93	91	96	84	85	98	90	66	94	89	88	85
DP SVR	93	89	95	87	75	98	90	65	84	83	84	77

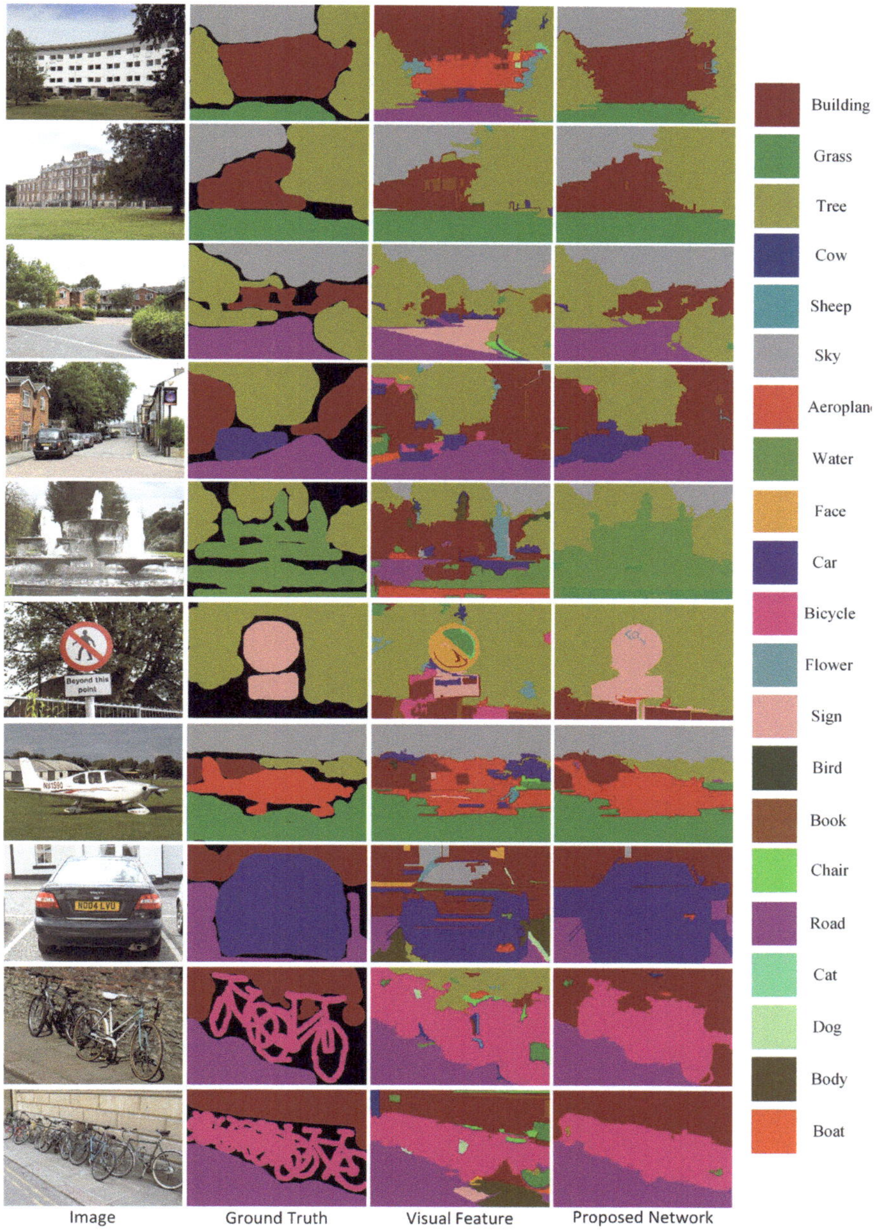

Fig. 4.13 Qualitative results on the MSRC dataset. The deep learning network using CAV features successfully removes a significant portion of misclassification in the results of visual feature based prediction

Table 4.9 Confusion matrix for 21 objects on the MSRC dataset (SVR, global accuracy 85.5%)

	Building	Grass	Tree	Cow	Sheep	Sky	Aero.	Water	Face	Car
Build	93.9	0.6	1.1	0.1	0.0	1.2	0.3	0.2	0	0.8
Grass	1.3	87.1	1.3	4.4	1.5	0.6	0.4	0.2	0	0.1
Tree	3.4	2.0	91.3	0.1	0	1.5	0.5	0.1	0	0.4
Cow	0.1	9.2	1.2	88.7	0.1	0.4	0	0	0	0
Sheep	0	10.3	0.1	0.3	89.0	0	0	0	0	0
Sky	11.4	0	2.0	0	0	85.9	0.2	0.1	0	0
Aero	6.0	4.3	1.0	0	0	2.1	82.6	0.2	0	0.5
Water	3.2	1.4	0.2	1.0	0	13.2	0.4	70.6	0	0.1
Face	2.5	0	1.5	0.1	0.1	0.6	0	0.1	81.9	0.1
Car	3.9	0.1	0.7	0	0	0	0.2	0.1	0	93.4
Bicycle	5.8	0.4	1.5	0	0	0.2	0	0.2	0.1	0.2
Flower	0.6	0.3	2.1	0	0	0.4	0	0	0.6	0
Sign	5.0	0.4	1.4	0.2	0	2.5	0.2	0.1	0.3	0.2
Bird	0.6	9.2	0.5	0.4	0.4	4.4	0.8	4.0	0	0.1
Book	0.7	0.2	0.1	0	0.1	0.1	0	0.1	0.2	0.1
Chair	1.1	3.3	0.3	0.2	0.1	0.5	0.1	0.2	0.2	0.1
Road	2.9	13.0	0.3	0.1	0.2	4.0	0.9	0.4	0	5.7
Cat	0.3	6.0	1.3	0.4	0.9	1.5	0.1	0.1	0.9	0.1
Dog	0.6	5.9	0.5	0.8	0.1	0.9	0	0.5	2.0	0
Body	2.2	1.6	1.7	0.1	0.5	2.3	0	0.2	3.6	0.8
Boat	1.7	0.4	0.3	0.8	0.1	0.6	3.6	13.7	0	0.1

(continued)

Table 4.9 (continued)

	Bicycle	Flower	Sign	Bird	Book	Chair	Road	Cat	Dog	Body	Boat
Build	0.5	0	0.2	0	0.3	0	0.5	0	0	0.1	0.1
Grass	0.2	0	0	0.5	0	0.1	1.5	0	0	0.8	0.1
Tree	0.5	0	0	0	0.1	0	0.1	0	0	0.1	0.1
Cow	0	0	0	0	0.1	0	0	0.1	0	0	0.1
Sheep	0.1	0	0	0	0.1	0	0.1	0.1	0	0	0
Sky	0	0	0.3	0	0	0	0	0	0	0	0
Aero	0	0	0.5	0	0	0	0.6	0	0	0.1	2.1
Water	0	0	0.1	2.2	0	0	2.7	0	0	0.8	4.2
Face	0.1	0.2	0.1	0	1.2	0	0.4	0.2	0.1	10.7	0
Car	0.2	0	0	0	0	0	1.4	0	0	0.1	0
Bicycle	88.7	0	0	0	0.1	0	2.7	0	0	0.1	0
Flower	0.3	95.0	0	0	0	0	0.1	0.2	0.1	0.1	0
Sign	0.8	0.2	86.7	0.2	0.4	0.1	0.3	0	0	0.3	0.6
Bird	1.1	0.1	0.3	74.9	0	0	1.0	1.5	0.1	0	0.8
Book	0.1	0	0.2	0	97.9	0	0.1	0	0.1	0.2	0
Chair	0.3	0.1	0.2	0.1	0.5	89.7	2.4	0	0.1	0.5	0.3
Road	1.4	0	0.1	0.3	0	2.0	65.3	0.7	2.3	0.5	0
Cat	0.1	1.5	0	0.4	0.1	0	2.3	83.6	0.4	0.2	0
Dog	0.2	0.2	0	0.1	0.1	0	3.8	0.9	82.8	0.5	0.2
Body	0.2	0.2	0	0.1	0.3	0.1	1.9	0.1	0.2	83.9	0.2
Boat	0.6	0	0.3	0	0	0	1.0	0	0	0.1	76.7

Table 4.10 Performance (%) comparisons with previous approaches on the SIFT Flow dataset

Refs.	Global acc.	Avg. acc.
Liu et al. [42]	74.75	–
Najafi et al. [57]	76.6	35.0
Nguyen et al. [58]	78.9	34.0
Shuai et al. [51]	80.1	39.7
Sharma et al. [35]	80.9	39.1
Visual feature (ANN)	67.7	17.0
DP network (ANN)	86.9	36.9
Visual feature (SVR)	63.4	16.2
DP network (SVR)	87.0	36.7

network also has one of the top class accuracies of 36.9%. The results indicate the advantages of incorporating CAV contextual features in overcoming the complexity of object classification in natural scenes. Figure 4.14 visually compares classification results on sample images between the DP network and visual feature based classifiers, from which we can observe the capacity of CAV features in removing misclassification.

4.5.4 Discussions

The main lessons that can be learnt from the experimental results are as follows:

(1) The CAV features yields dramatic improvements in the performance compared with using visual features alone. This is primarily due to the merit of CAV features in capturing both long- and short-range label dependencies between objects, being able to adapt to the local properties of test images, and preserving both relative and absolute location information. The CAV features reflect a rich set of semantic contextual information embedded in each type of scene, and thus are useful for enforcing contextual constraints on class labels of complex objects.

(2) The deep learning network shows higher accuracy than or is competitive to state-of-the-art scene parsing systems on three benchmark datasets, although its system parameters are simply pre-set and have not been optimized. Its architecture is also general, as its three layers are relatively independent and can directly adopt various existing algorithms. For instance, the classifier in the first layer can be chosen from different probabilistic prediction models, such as ANN, SVR and Adaboost.

(3) The deep learning network significantly improves the recorded accuracy from 80.9 to 87% on the SIFT Flow dataset. This is largely due to a large number of training images in the dataset, which can ensure collecting sufficient object distribution information for generating reliable CAV features. Thus, the

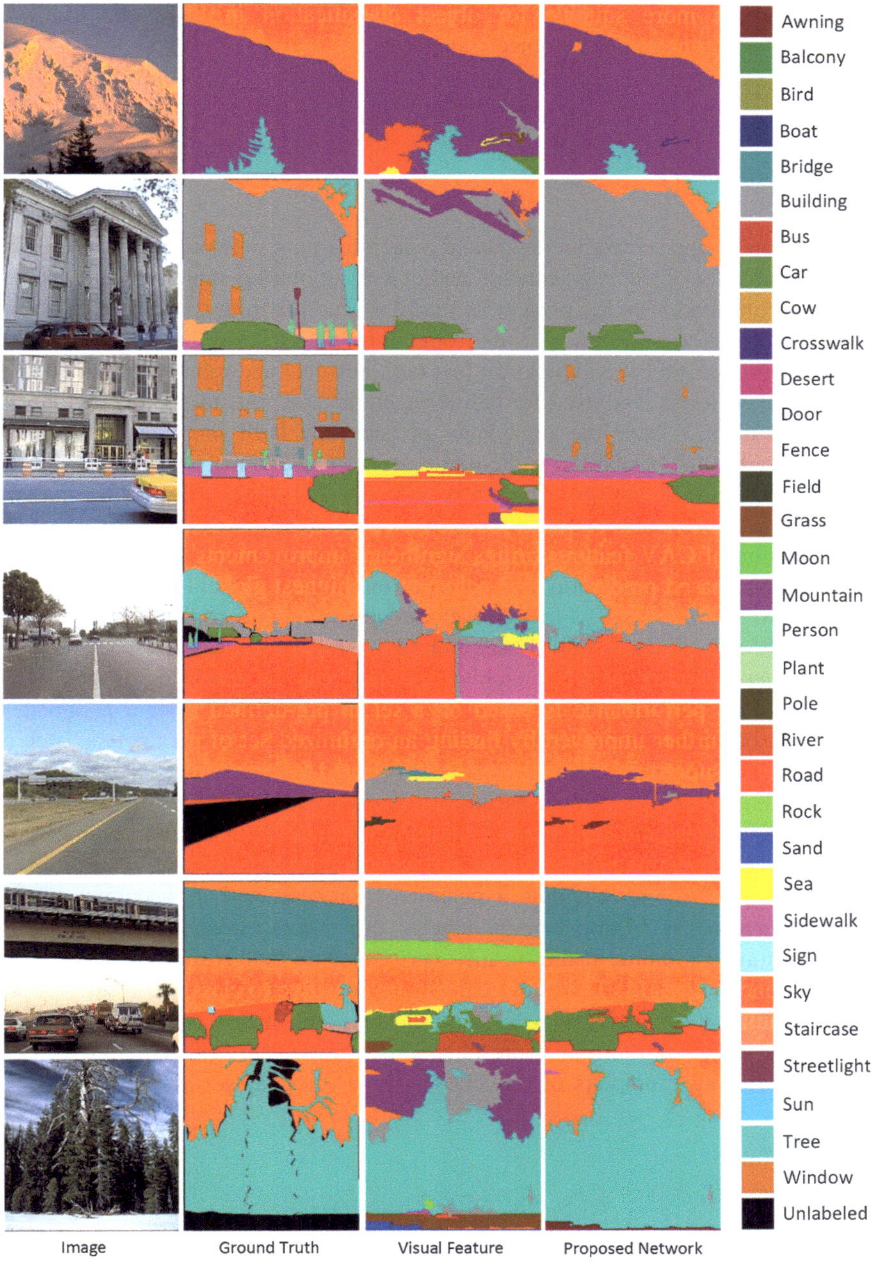

Fig. 4.14 Qualitative results on the SIFT Flow dataset. The results demonstrate the benefit of incorporating CAV features to improve the classification accuracy and correct misclassification error for complex objects in natural scenes

network is more suitable for object classification in large datasets with pixel-wise annotations of object categories.

4.5.5 Summary

This section has presented and evaluated a deep learning network architecture with CAV features for object segmentation and classification in complex natural images. One key contribution to the present literature is incorporating spatial image blocks to collect object co-occurrence priors from the training data by pooling over object pixel distributions between blocks so that both absolute and relative coordinates of objects are preserved and used for enforcing contextual constraints on class labels. The CAV features have the advantages of capturing both short- and long-range label dependencies of objects in the entire scene, while being able to adapt to the local properties in the scene. Thus, they are very effective to remove misclassification error in the results of visual feature based prediction. It is also shown that the incorporation of CAV features brings significant improvements to the accuracy of visual feature based prediction, and achieves the highest global accuracy of 87.0% recorded so far on the SIFT Flow dataset and very competitive global accuracies of 81.0 and 85.0% on the Stanford background and MSRC datasets respectively.

The deep learning network architecture still can be extended from several aspects: (1) its performance is based on a set of pre-defined system parameters, which can be further improved by finding an optimized set of parameters, such as the number of superpixels, the dimension of selected feature subsets, and the length of ANN hidden neurons. (2) It assumes a linear relationship between visual and contextual features, and it is worth exploiting the modelling of more complicated non-linear relationships between them by incorporating techniques, such as logistic regression models. (3) It adopts a parametric process, and its accuracy can be further increased by adding a nonparametric pro-processing step, which retrieves the most similar training images to the query image to collect more reliable object co-occurrence priors. (4) The CAV features are sensitive to class pixel distributions in the training data, particularly for rare classes, and thus it is still necessary to investigate strategies that pay more attention to those rare but important classes, such as enriching exemplars of rare classes from other datasets.

References

1. L. Zheng, Y. Zhao, S. Wang, J. Wang, Q. Tian, Good practice in CNN feature transfer. *arXiv preprint* arXiv:1604.00133 (2016)
2. S.D. Learning, CS231n: convolutional neural networks for visual recognition (2016). http://cs231n.github.io/convolutional-networks/

3. J. Ba, V. Mnih, K. Kavukcuoglu, Multiple object recognition with visual attention, *arXiv preprint* arXiv:1412.7755 (2014)
4. J. Donahue, L. Anne Hendricks, S. Guadarrama, M. Rohrbach, S. Venugopalan, et al., Long-term recurrent convolutional networks for visual recognition and description, in *Computer Vision and Pattern Recognition (CVPR), IEEE Conference on* (2015), pp. 2625–2634
5. A. Dundar, J. Jin, E. Culurciello, Convolutional clustering for unsupervised learning. *arXiv preprint* arXiv:1511.06241 (2015)
6. D.V. Nguyen, L. Kuhnert, K.D. Kuhnert, Structure overview of vegetation detection. A novel approach for efficient vegetation detection using an active lighting system. Robot. Auton. Syst. **60**, 498–508 (2012)
7. I. Lenz, H. Lee, A. Saxena, Deep learning for detecting robotic grasps. Int. J. Robot. Res. **34**, 705–724 (2015)
8. L. Romaszko, A deep learning approach with an ensemble-based neural network classifier for black box ICML 2013 contest, in *Workshop on Challenges in Representation Learning, International Conference on Machine Learning (ICML)* (2013), pp. 1–3
9. S. Ahmad Radzi, K.-H. Mohamad, S.S. Liew, R. Bakhteri, Convolutional neural network for face recognition with pose and illumination variation. Int. J. Eng. Technol. (IJET) **6**, 44–57 (2014)
10. F. Shaheen, B. Verma, M. Asafuddoula, Impact of automatic feature extraction in deep learning architecture, in *Digital Image Computing: Techniques and Applications (DICTA), International Conference on* (2016), pp. 1–8
11. C. Cortes, Y. LeCun, C.J.C. Burges, The MNIST database of handwritten digits. http://yann.lecun.com/exdb/mnist/
12. K. He, X. Zhang, S. Ren, J. Sun, Deep residual learning for image recognition. *arXiv preprint* arXiv:1512.03385 (2015)
13. Y. Lecun, L. Bottou, Y. Bengio, P. Haffner, Gradient-based learning applied to document recognition. Proc. IEEE **86**, 2278–2324 (1998)
14. D.V. Nguyen, L. Kuhnert, K.D. Kuhnert, Spreading algorithm for efficient vegetation detection in cluttered outdoor environments. Robot. Auton. Syst. **60**, 1498–1507 (2012)
15. D.V. Nguyen, L. Kuhnert, T. Jiang, S. Thamke, K.D. Kuhnert, Vegetation detection for outdoor automobile guidance, in *Industrial Technology (ICIT), IEEE International Conference on* (2011), pp. 358–364
16. A. Bosch, X. Muñoz, J. Freixenet, Segmentation and description of natural outdoor scenes. Image Vis. Comput. **25**, 727–740 (2007)
17. W. Guo, U.K. Rage, S. Ninomiya, Illumination invariant segmentation of vegetation for time series wheat images based on decision tree model. Comput. Electron. Agri. **96**, 58–66 (2013)
18. F. Shaheen, B. Verma, An ensemble of deep learning architectures for automatic feature extraction, in *Computational Intelligence (ISSCI), IEEE Symposium Series on* (2016) (in Press)
19. D.-X. Liu, T. Wu, B. Dai, Fusing ladar and color image for detection grass off-road scenario, in *Vehicular Electronics and Safety (ICVES), IEEE International Conference on* (2007), pp. 1–4
20. R. Mottaghi, S. Fidler, A. Yuille, R. Urtasun, D. Parikh, Human-machine CRFS for identifying bottlenecks in scene understanding. Pattern Anal. Mach. Intell. IEEE Trans. **38**, 74–87 (2016)
21. J. Shotton, J. Winn, C. Rother, A. Criminisi, Textonboost for image understanding: multi-class object recognition and segmentation by jointly modeling texture, layout, and context. Int. J. Comput. Vis. **81**, 2–23 (2009)
22. S. Gould, J. Rodgers, D. Cohen, G. Elidan, D. Koller, Multi-class segmentation with relative location prior. Int. J. Comput. Vis. **80**, 300–316 (2008)
23. Y. Jimei, B. Price, S. Cohen, Y. Ming-Hsuan, Context driven scene parsing with attention to rare classes, in *Computer Vision and Pattern Recognition (CVPR), IEEE Conference on* (2014), pp. 3294–3301

24. A. Singhal, L. Jiebo, Z. Weiyu, Probabilistic spatial context models for scene content understanding, in *Computer Vision and Pattern Recognition, (CVPR), IEEE Conference on* (2003), pp. 235–241
25. B. Micusik, J. Kosecka, Semantic segmentation of street scenes by superpixel co-occurrence and 3D geometry, in *Computer Vision Workshops (ICCV Workshops), IEEE 12th International Conference on* (2009), pp. 625–632
26. C. Farabet, C. Couprie, L. Najman, Y. LeCun, Learning hierarchical features for scene labeling. Pattern Anal. Mach. Intell. IEEE Trans. **35**, 1915–1929 (2013)
27. M. Seyedhosseini, T. Tasdizen, Semantic image segmentation with contextual hierarchical models. Pattern Anal. Mach. Intell. IEEE Trans. **38**(5), 951–964 (2015)
28. D. Batra, R. Sukthankar, C. Tsuhan, Learning class-specific affinities for image labelling, in *Computer Vision and Pattern Recognition, (CVPR), IEEE Conference on* (2008), pp. 1–8
29. Z. Lei, J. Qiang, Image segmentation with a unified graphical model. Pattern Anal. Mach. Intell. IEEE Trans. **32**, 1406–1425 (2010)
30. R. Xiaofeng, B. Liefeng, D. Fox, RGB-(D) scene labeling: features and algorithms, in *Computer Vision and Pattern Recognition (CVPR), IEEE Conference on* (2012), pp. 2759–2766
31. A.G. Schwing, R. Urtasun, Fully connected deep structured networks. *arXiv preprint* arXiv:1503.02351 (2015)
32. S. Zheng, S. Jayasumana, B. Romera-Paredes, V. Vineet, Z. Su, et al., Conditional random fields as recurrent neural networks. *arXiv preprint* arXiv:1502.03240 (2015)
33. P.H. Pinheiro, R. Collobert, Recurrent convolutional neural networks for scene parsing. *arXiv preprint* arXiv:1306.2795 (2013)
34. A. Sharma, O. Tuzel, M.-Y. Liu, Recursive context propagation network for semantic scene labeling, in *Advances in Neural Information Processing Systems* (2014), pp. 2447–2455
35. A. Sharma, O. Tuzel, D.W. Jacobs, Deep hierarchical parsing for semantic segmentation, in *Computer Vision and Pattern Recognition (CVPR), IEEE Conference on* (2015), pp. 530–538
36. S. Ling, L. Li, L. Xuelong, Feature learning for image classification via multiobjective genetic programming. Neural Netw. Learn. Syst. IEEE Trans. **25**, 1359–1371 (2014)
37. L. Zhang, B. Verma, D. Stockwell, S. Chowdhury, Spatially constrained location prior for scene parsing, in *Neural Networks (IJCNN), International Joint Conference on* (2016), pp. 1480–1486
38. J. Tighe, S. Lazebnik, Superparsing: scalable nonparametric image parsing with superpixels, in *Computer Vision (ECCV), European Conference on* (2010), pp. 352–365
39. P. Hanchuan, L. Fuhui, C. Ding, Feature selection based on mutual information criteria of max-dependency, max-relevance, and min-redundancy. Pattern Anal. Mach. Intell. IEEE Trans. **27**, 1226–1238 (2005)
40. P. Felzenszwalb, D. Huttenlocher, Efficient graph-based image segmentation. Int. J. Comput. Vis. **59**, 167–181 (2004)
41. S. Gould, R. Fulton, D. Koller, Decomposing a scene into geometric and semantically consistent regions, in *Computer Vision (ICCV), IEEE 12th International Conference on* (2009), pp. 1–8
42. L. Ce, J. Yuen, A. Torralba, Nonparametric scene parsing: label transfer via dense scene alignment, in *Computer Vision and Pattern Recognition (CVPR), IEEE Conference on* (2009), pp. 1972–1979
43. V. Lempitsky, A. Vedaldi, A. Zisserman, Pylon model for semantic segmentation, in *Advances in Neural Information Processing Systems* (2011), pp. 1485–1493
44. D. Munoz, J.A. Bagnell, M. Hebert, Stacked hierarchical labeling, in *Computer Vision (ECCV), European Conference on* (2010), pp. 57–70
45. L. Ladicky, C. Russell, P. Kohli, P.H.S. Torr, Associative hierarchical random fields. Pattern Anal. Mach. Intell. IEEE Trans. **36**, 1056–1077 (2014)
46. A. Krizhevsky, I. Sutskever, G.E. Hinton, ImageNet classification with deep convolutional neural networks, in *Advances in Neural Information Processing Systems* (2012), pp. 1097–1105

47. M.D. Zeiler, R. Fergus, Visualizing and understanding convolutional networks, in *European Conference on Computer Vision* (2014), pp. 818–833
48. C. Szegedy, W. Liu, Y. Jia, P. Sermanet, S. Reed, et al., Going deeper with convolutions, in *Computer Vision and Pattern Recognition (CVPR), IEEE Conference on* (2015), pp. 1–9
49. K. Simonyan, A. Zisserman, Very deep convolutional networks for large-scale image recognition. *arXiv preprint* arXiv:1409.1556 (2014)
50. K. He, X. Zhang, S. Ren, J. Sun, Identity mappings in deep residual networks. *arXiv preprint* arXiv:1603.05027 (2016)
51. S. Bing, W. Gang, Z. Zhen, W. Bing, Z. Lifan, Integrating parametric and non-parametric models for scene labeling, in *Computer Vision and Pattern Recognition (CVPR), IEEE Conference on* (2015), pp. 4249–4258
52. J. Shotton, J. Winn, C. Rother, A. Criminisi, Textonboost: joint appearance, shape and context modeling for multi-class object recognition and segmentation, in *Computer Vision (ECCV), European Conference on* (2006), pp. 1–15
53. Z. Long, C. Yuanhao, L. Yuan, L. Chenxi, A. Yuille, Recursive segmentation and recognition templates for image parsing. Pattern Anal. Mach. Intell. IEEE Trans. **34**, 359–371 (2012)
54. E. Akbas, N. Ahuja, Low-level hierarchical multiscale segmentation statistics of natural images. Pattern Anal. Mach. Intell. IEEE Trans. **36**, 1900–1906 (2014)
55. A. Lucchi, L. Yunpeng, P. Fua, Learning for structured prediction using approximate subgradient descent with working sets, in *Computer Vision and Pattern Recognition (CVPR), IEEE Conference on* (2013), pp. 1987–1994
56. C. Gatta, F. Ciompi, Stacked sequential scale-spacetaylor context. Pattern Anal. Mach. Intell. IEEE Trans. **36**, 1694–1700 (2014)
57. M. Najafi, S.T. Namin, M. Salzmann, L. Petersson, Sample and filter: nonparametric scene parsing via efficient filtering. *arXiv preprint* arXiv:1511.04960 (2015)
58. T.V. Nguyen, L. Canyi, J. Sepulveda, Y. Shuicheng, Adaptive nonparametric image parsing. Circ. Syst. Video Technol. IEEE Trans. **25**, 1565–1575 (2015)

Chapter 5
Case Study: Roadside Video Data Analysis for Fire Risk Assessment

In this chapter, we present a case study of utilizing machine learning techniques for fire risk assessments on roadside video data.

5.1 Introduction

Accurately estimating the site-specific parameters of roadside grasses such as biomass, height, coverage and density plays an important role in many applications such as assisting growth condition monitoring and roadside vegetation management. These parameters can provide reliable and important indications of the current grass condition, growth stage and future tendency. Tracking the changes in these parameters is an effective way to detect and quantify the state of the vegetation such as diseases, dryness, soil nutrients and water stress. For road drivers and vehicles, vegetation with high biomass can present a major fire hazard to their security, particularly in remote regions without regular and frequent human checks on roadside grass growth conditions. The development of automatic and efficient methods to estimate the biomass of roadside grasses is thus very important for the transport authority identifying fire-prone road regions and taking necessary actions to burn or cut the grasses to prevent possible hazards.

Biomass is typically defined as the dry mass of the above ground portion of vegetation [1]. Studies have found [2–5] that there is a close statistical relationship between the plant height and biomass yield, although this relationship may depend on the type of vegetation. Existing approaches to the estimation of vegetation biomass can be roughly categorized into three groups: (a) conducting field surveys, which involve destructive samplings of plants at different growth stages, counting of the number of plants contained in the sample and calculating the weighting after drying them [6]. However, this method often requires a considerable amount of investment in terms of time, labour, and cost, and is infeasible for a large-scale field survey. (b) Most existing studies [7] are focusing on remote sensing methods,

B. Verma et al., *Roadside Video Data Analysis: Deep Learning*,
Studies in Computational Intelligence 711, DOI 10.1007/978-981-10-4539-4_5

which enable assessment of vegetation characteristics including biomass and height using data collected with various optical imaging sensors mounted on space-borne, airborne, and terrestrial platforms. However, remote sensing methods often focus on a large-scale vegetation field which has difficulty in site-specific analysis, and they also suffer from high expense and are prone to be impacted by atmospheric conditions such as rain and cloud cover. (c) Recently, there are a few studies [8–12] that have investigated the measurement of tree and rice plant height using image processing techniques on ground-based data. The plant height is measured based on the distance between pre-set reference markers. These approaches often require proper settings of data capturing equipment with respect to the location, angle, height, etc. and manual assistance to install reference markers, and thus they have very limited applicability for real-world applications.

5.2 Related Work

Related work can be approximately grouped into three categories—human field survey, remote sensing measurement, and image processing technique.

(1) A traditional way of estimating the plant height is to conduct field surveys and visually inspect the height by humans, which is generally of high accuracy but demands time- and labour-intensive human efforts and large costs. Field surveys may also have the difficulty accessing some areas due to geographical conditions or the requirement of permission from private land owners or relevant authorities.
(2) Most of existing automatic analysis systems are heavily dependent on remote sensing measurements. One type of feature normally used is the VI. Payero et al. [13] compared 11 VIs for estimating the plant height of two crops—grass and alfalfa, and found that only 4 indices have a good linear relationship with the height. They recommended that an appropriate VI should be selected for a particular type of crop and for a specific height being considered. Since the 1980s, studies have been focusing more on designing measurement methods based on data from airborne LIDAR sensors [1], which provide more accuracy and detailed information about the canopy than others such as VIs and ultrasonic sensors [14]. The canopy height model is determined by taking the difference between the digital surface model, which represents the elevation of the topmost layers of the forest canopy, and the Digital Terrain Model (DTM), which indicates continuous elevation of the ground surface [15, 16]. To alleviate the requirement of a DTM, Yamamoto et al. [17] measured the mean tree height as the difference between the ground return and a top surface model, which is nearly parallel to a DTM, achieving nearly 1-m accuracy. In [18], a stereo matching algorithm was proposed to estimate the height of objects from monocular aerial images. A review of existing studies on estimating plant

biomass using LIDAR data can be found in [7]. However, studies using data from satellite or airplane often focus on large-scale vegetation fields which have difficulty supporting site-specific analysis, suffer from high expense, and are prone to be impacted by atmospheric conditions such as rain and cloud weathers. Recently, ground-based laser scanning technology has also been used to estimate site-specific crop heights. In [19], the stem height of a crop *Miscanthus giganteus* was measured by the average of the combined LIDAR visible above-ground stem lengths. In [4], the plant height of paddy rice was calculated using crop surface models generated from the terrestrial laser scanning-derived point clouds. Due to the requirement of setting-up LIDAR equipment, these approaches support only site-specific applications, and have limited applicability in large-scale fields.

(3) The estimation of the plant height from ground-based data using machine learning techniques is a largely unexplored field. In [8], an approach for detecting the rice plant height from continuously monitored daily photos of a rice crop field was proposed. To provide a height reference, a marker bar of a known height was installed in the field and the height of rice was obtained by comparing the height of the rice with the height of the marker bar. In a similar approach for measuring the height of trees [9], two red marker points were pre-set on the root and one meter distance from the root of a tree, respectively and the height of a tree was calculated based on a proportion transform of the coordinates of maker points and the top point of the tree. This approach was later deployed in a mobile phone platform [10], and further extended by including an additional marker point and using a perspective transformation [11]. These approaches actually base the task of determining the plant height on a process of segmenting the reference markers using image processing techniques. They require a strict field setting with respect to the height, location, and angle of data capturing equipment, as well as the location and visibility of reference markers. The techniques used are workable only for a specific site and do not support automatic large-scale field analysis.

Overcoming drawbacks in existing studies, the Vertical Orientation Connectivity of Grass Pixels (VOCGP) approach presented in the case study is built for ground-based images collected using ordinal digital cameras rather than satellite or airplane data. It is fully automatic and easy to operate, and does not require manually setting up height reference markers, nor the availability of any specified equipment. It supports not only site-specific analysis, but also large-scale field tests. One pre-requisite of the approach is that the distance of the camera to grasses should be approximately fixed, which can be controlled during data collection. Thus, the approach is automatic and efficient, and has high flexibility and applicability.

To validate different impacts of using deep and non-deep learning algorithms on the calculation results of VOGCP, we include two classifiers—ANN and CNN for grass region segmentation and compare the prediction accuracies of them on the same dataset.

5.3 Proposed VOCGP Approach

5.3.1 Problem Formulation and Motivations

The classical way of measuring grass biomass in field surveys often involves destructive samplings of grass stems in a sampling region, and calculating their weights. This way of calculating the fuel load of grasses (tonnes/ha) can be mathematically expressed based on the properties of grass stems:

$$\mathrm{F}_l = \frac{1}{N_s}\sum_{j=1}^{N_s} s_j * u_j \tag{5.1}$$

where, s_j stands for the length of the jth stem, u_s is a fuel load unit (e.g. fuel load per meter of grasses) for the jth stem, N_s is the total number of stems. The fuel loads are averaged over all stems.

Let $W = \{X_1, X_2, \ldots, X_i, \ldots, X_D\}$ be the corresponding sampling grass window in an image, $X_i = \{x_{i1}, x_{i2}, \ldots, x_{ij}, \ldots, x_{iH}\}$ indicates the ith column vector of W, $W \in R^{H*D}$, H and D represent the number of rows and columns respectively, the objective of grass biomass estimation is to seek a mapping function that projects W to an estimated fuel load F_w:

$$\mathrm{F}_w = f(W) \tag{5.2}$$

so that the difference between estimated and physically quantified fuel loads can be minimized:

$$\mathrm{F} = \min\left(\mathrm{F}_w - \mathrm{F}_l\right) \tag{5.3}$$

Assume the grass biomass is closely correlated with grass height and density. To simulate the way of calculating the fuel load of grasses in field surveys using (5.1), the grass stems can be regarded as being approximately represented by column vectors in images. Thus, the estimated fuel load in (5.2) can be calculated based on the lengths of grass pixels in all columns in W:

$$\mathrm{F}_w = f(W) = \frac{1}{D}\sum_{i=1}^{D} l_i \times f_i \tag{5.4}$$

where, l_i is the length of grass pixels in the ith column X_i,f_i is a calibration factor that makes l_i directly comparable with the fuel load $s_j \times u_j$, and D is the total number of columns in W.

Equation 5.4 enables us to estimate the fuel load of grasses in images using a similar concept to that in field surveys, and it forms the basic idea of the VOCGP approach for estimating the grass biomass, which measures the grass height and density using the average connectivity of grass pixels in a vertical orientation in all columns of a sampling window. Without losing generality, the function in (5.4) can be accomplished in two separate tasks: finding a proper length measurement l_i for the ith column and combining the lengths of all columns:

$$l_i = f_1(X_i) \tag{5.5}$$

$$\mathrm{F}_w = f_2\left(\bigcup_{i=1,\ldots,D} l_i\right) \tag{5.6}$$

The above two equations will be solved in the VOCGP approach. The VOCGP is designed based on the finding that there is a statically close relationship between the plant height and biomass yield [2–5]. However, we believe that using only height may not be robust to the scenarios where low and high grasses coexist in the sampling region. Thus, we also include the grass density as an additional factor in the design of VOCGP. It is generally more confident that the higher height and higher density of grasses, the higher biomass. In measurement of grass height and density, we are inspired by previous work [20–22] which showed that the dominant texture orientation at each image pixel can be used to generate a robust indicator of the primary direction in a large neighbourhood. We observed that high and density grasses often have long unbroken pixel connectivity along the vertical orientation, while low and sparse grasses normally have short and broken connectivity in most columns of the sampling window.

5.3.2 *Overview of Approach*

As shown in Fig. 5.1, the VOCGP approach consists of four major steps: (1) sampling window selection, (2) grass region segmentation, (3) vertical orientation detection, and (4) VOCGP calculation.

A sampling grass window is firstly selected from an input image which is taken as the basic processing unit for estimating the biomass. The reason is that the total biomass produced by the grasses is sensitive to the size and location of the region from which the grass stems are destructively sampled and measured. We follow the practice that the sampling region is often set to the same size for all sampling sites, and from a location of a particular interest in each site [23].

Within a sampling window, two classifiers are compared for grass region segmentation: ANN and CNN. For ANN, color and texture features that represent discriminative characteristics of grass vs. non-grass pixels are extracted and a feature-level fusion of them is further fed into the ANN classifier. For the CNN classifier, the pixel values in the original window are used. As a parallel processing, the dominant texture orientation at each pixel is also detected by performing votes on the responses of multi-resolution and multi-scale Gabor filters. Based on

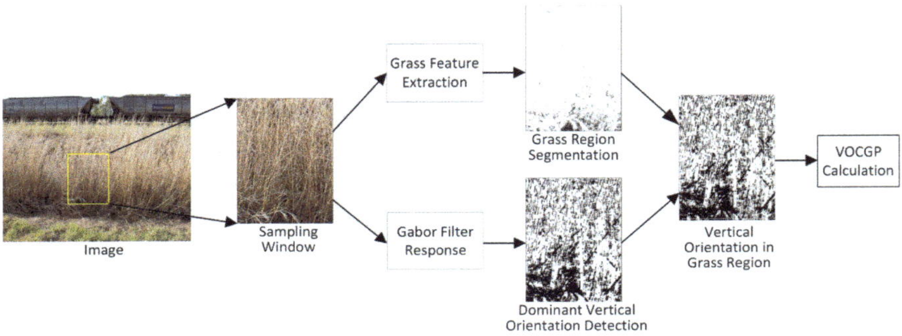

Fig. 5.1 Graphic illustration of main processing steps in the VOCGP approach. For a given image, the approach outputs a VOCGP value for estimating the grass biomass within a sampling window. The outputs of grass region segmentation and dominant vertical orientation detection are combined for VOCGP calculation

segmented grass results and dominant vertical orientations, we then propose an algorithm to calculate the VOCGP, which is the mean length of grass pixels with a continuously connected dominant vertical orientation in all columns of the window. The VOCGP is used as an estimator of the biomass of grasses within the window.

5.3.3 Grass Region Segmentation

Grass region segmentation aims to distinguish grass pixels from non-grass pixels. The outputs can provide an indication of the spatial distribution and coverage of grass pixels within the sampling window. Two types of techniques are comparatively used for grass region segmentation: (1) ANN classifier with human engineering grass features, and (2) CNN with automatic feature extraction.

(1) ANN classifier with grass features. An effective representation of the visual characteristics in grass pixels plays a critical role in grass region segmentation. The first technique generates color and texture fused features for roadside grass analysis. It is well known that grasses are primarily represented in a green or yellow color, and have a richer unstructured texture such as edges than other objects such as sky and road. Fusion of them is expected to lead to more accurate and robust segmentation results.

    The color spaces include CIELab and RGB. The Lab was found perceptually consistent with the human eyes [24], and RGB may contain complementary information for grasses that Lab fails to provide. The texture includes features from 17-D filter banks firstly adopted in [25], which have shown high discriminative capacity for texture-rich objects and been widely adopted for generic object classification [26]. The 17-D filter banks include Gaussians with 3 different scales (1, 2, 4) applied to Lab channels, Lapacians of Gaussians with 4 different scales (1, 2, 4, 8) and the derivatives of Gaussians with two different scales (2, 4) for each axis (x and y) on the L channel.

The color and texture feature vectors of a pixel at coordinates (i, j) are composed of:

$$V_{i,j}^{c} = [R, G, B, L, a, b] \tag{5.7}$$

$$V_{i,j}^{t} = \left[G_{1,2,4}^{L}, G_{1,2,4}^{a}, G_{1,2,4}^{b}, LOG_{1,2,4,8}^{L}, DOG_{2,4,x}^{L} DOG_{2,4,y}^{L}\right] \tag{5.8}$$

They are fused to obtain a 23-element color and texture feature vector:

$$V_{i,j} = \left[V_{i,j}^{c},\ V_{i,j}^{t}\right] \tag{5.9}$$

We incorporate a binary ANN classifier for discriminating grass vs. non-grass pixels based on color and texture features. The ANN accepts an input feature vector $V_{i,j}$ and outputs probabilities for two classes:

$$p_{i,j}^{k} = tran\,(w_k V_{i,j} + b_k) \tag{5.10}$$

where, *tran* stands for the prediction function of a three-layer transig/purelin ANN, and w_k and b_k are the trained weights and constant parameters for the kth object class. The class having the highest probability across all classes wins the classification label:

$$A_{i,j} = \max_{k \in C} p_{i,j}^{k} \tag{5.11}$$

where, C stands for grass and non-grass classes, and $A_{i,j}$ represents a binary label (1—grass, and 0—non-grass) for a pixel at (i, j).

(2) CNN classifier with automatic feature extraction. CNNs accept the raw image pixels as inputs, and utilize convolutional and pooling layers to progressively extract more abstract patterns, which are further input into a full connection layer to generate predictions of object categories. The popular LeNet-5 CNN [27] is adopted. For a pixel at (i, j) in the sampling window W, the LeNet-5 provides a decision on a binary category $A_{i,j} \in \{grass, non - grass\}$.

5.3.4 Gabor Filter Voting for Dominant Vertical Orientation Detection

This part presents a Gabor filter based voting approach to detect the dominant vertical orientation at each image pixel. The precise estimate of the dominant texture orientation at each image pixel is crucial in the determination of grass height and density. The biologically inspired Gabor filters are one of the most popular multi-resolution texture descriptors for representing and discriminating the appearance properties of objects. They are useful for extracting a rich set of multi-scale and orientation features such as edge, line, and structured texture, for pattern analysis. We perform votes on Gabor filter responses in multiple orientations to detect the strongest local texture orientation at each pixel by taking into consideration all pixel intensities in a small spatial neighbourhood.

The 2D Gabor filter function [28] can be mathematically expressed as:

$$F(x,y) = \frac{f^2}{\pi\gamma\eta}\exp\left(-\left(\frac{f^2}{\gamma^2}X^2 + \frac{f^2}{\eta^2}Y^2\right)\right)\exp(j2\pi fX) \tag{5.12}$$

$$X = x\cos\theta + y\sin\theta \text{ and } Y = -x\sin\theta + y\cos\theta \tag{5.13}$$

where, (x, y) defines the centre of the filter, f indicates the central frequency of the filter, θ is the orientation, γ and η are the sharpness among the Gaussian major and minor axes perpendicular to the wave respectively. The spatial aspect ratio is η/γ.

The frequency corresponds to scale information which can be calculated by

$$\phi_m = f_{max} \times k^{-m}, \quad m = \{0, 1, \ldots, M_\phi - 1\} \tag{5.14}$$

where ϕ_m indicates the mth scale, $f_0 = f_{max}$ is the highest frequency desired, $k > 1$ is the frequency scaling factor, and M_ϕ is the total number of scales.

The multiple orientations can be obtained using

$$\theta_n = 2n\pi/N_\theta, \quad n = \{0, 1, \ldots, N_\theta - 1\} \tag{5.15}$$

where, θ_n is the nth orientation and N_θ is the total number of orientations.

For the sampling widow W in RGB space, it is firstly converted to a grey scale by averaging R, G and B values of each pixel. The responses of a Gabor filter $F_{\theta,\phi}$ at an orientation θ and a scale ϕ can be obtained by convolving the filter with all pixels in W:

$$G_{\theta,\phi} = W \oplus F_{\theta,\phi} \tag{5.16}$$

The output $G_{\theta,\phi}$ is a complex value comprising of a real and an imaginary component. The two components are combined in a square norm to produce the complex magnitude which indicates the absolute response strength of Gabor filters:

$$\overline{G}_{\theta,\phi} = \sqrt{Real\,(G_{\theta,\phi})^2 + Img\,(G_{\theta,\phi})^2} \tag{5.17}$$

As we only care about the orientation information, the responses are then averaged over all scales to obtain a single response value for each orientation:

$$\overline{G}_\theta = \frac{1}{M_\phi} \sum_{m=0}^{M_\phi - 1} \overline{G}_{\theta,\phi_m} \tag{5.18}$$

where M_ϕ is the number of all scales. Thus, for a pixel at (i, j), an orientation vector that includes the response magnitudes along all orientations can be obtained:

$$\overline{G}_{i,j} = \left[\overline{G}_0^{i,j}, \overline{G}_1^{i,j}, \ldots, \overline{G}_{N_\theta - 1}^{i,j}\right] \tag{5.19}$$

where, N_θ indicates the number of all orientations. The dominant orientation of the pixel at (i, j) can be obtained by performing votes on responses along all orientations and taking the maximum one:

$$O_{i,j} = k \quad if \quad \overline{G}_n^{i,j} = \max\left(\overline{G}_{i,j}\right) = \max_{k=0,\ldots,N_\theta - 1} \overline{G}_k^{i,j} \tag{5.20}$$

For five-scale and four-orientation Gabor filters (i.e. $M_\phi = 5$; $N_\theta = 4$), Eq. 5.20 outputs an integer from [0, 1, 2, 3] for each pixel, indicating the dominant orientation of 0°, 45°, 90°, or 135° respectively as shown in Fig. 5.2. Because we consider only the vertical orientation 90°, i.e. $O_{i,j} = 2$, the output of this step is a decision on a vertical or non-vertical orientation at each pixel in the sampling window.

5.3.5 *Vertical Orientation Connectivity of Grass Pixels Calculation*

This part presents the algorithm of calculating the VOCGP value for a sampling grass window, which is an indicator of the grass height and density, and used as an

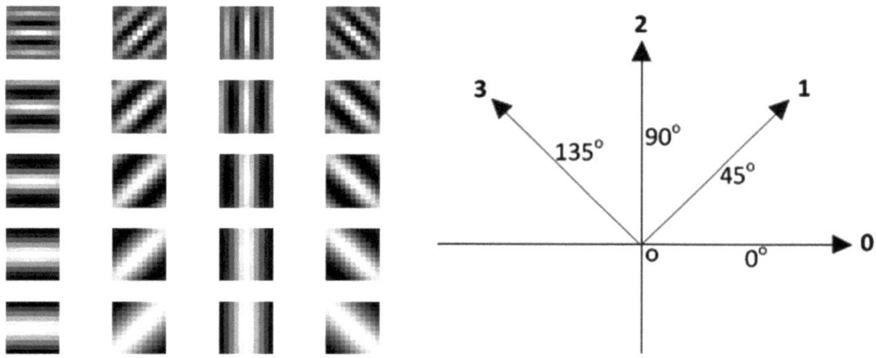

Fig. 5.2 Visual illustration of the real part of image responses using five-scale (*row*) and four-orientation (*column*) Gabor filters (*left*). Four orientations used in this case study and their defined indices (*right*)

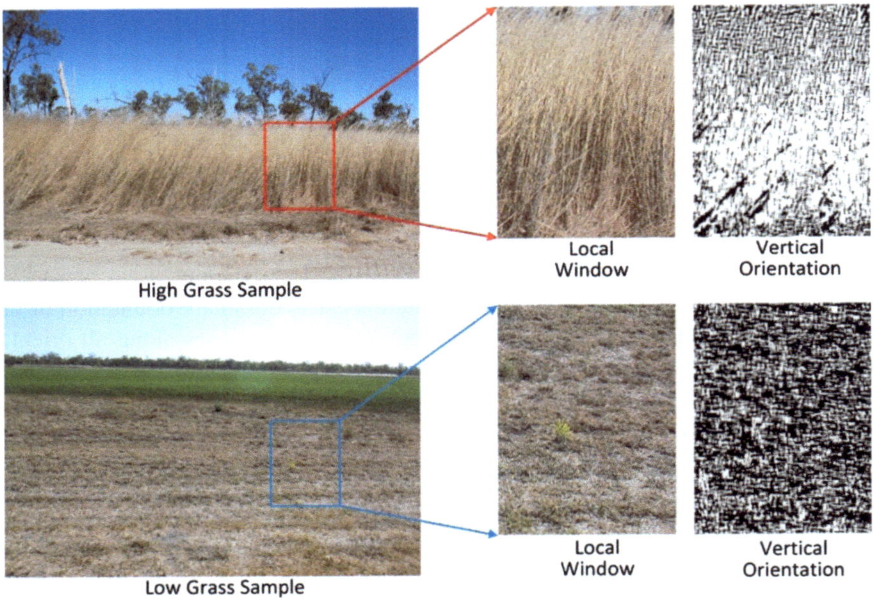

Fig. 5.3 Illustration of the difference in the connectivity along the vertical orientation between high and low grasses. The *white* and *black* in the right windows indicate the dominant vertical and non-vertical orientations respectively computed using Gabor filter votes. High grasses have longer unbroken connectivity along the vertical orientation than low grasses

estimator of the grass biomass. The VOCGP is defined as the mean length of continuously connected grass pixels with the same dominant vertical orientation in the window, and it is defined based on the observation that high grasses often have long unbroken pixel connectivity along the vertical orientation, while lower grasses normally have short and broken connectivity as illustrated in Fig. 5.3. In addition, dense grasses will have high connectivity in most columns of the window, while sparse grasses have high connectivity in only a limited number of columns.

From the segmented grass region $A_{i,j} \in \{\text{grass or non-grass}\}$ and detected dominant orientation $O_{i,j} \in \{\text{vertical or non-vertical}\}$, all pixels x_{ij} in the sampling window W are converted into a binary value which indicates whether the pixel belongs to the grass class and has a dominant vertical orientation:

$$x_{ij} = \begin{cases} 1, & \mathit{if}\ A_{i,j} = \text{grass}\ \mathit{and}\ O_{i,j} = \text{vertical} \\ 0, & \mathit{others} \end{cases} \tag{5.21}$$

For the ith column X_i, we calculate all lengths of continuously connected pixels x_{ij} with a value of 1:

$$C^i = \left\{c_1^i, c_2^i, \ldots, c_Q^i\right\} \quad (5.22)$$

where, c_q^i stands for the qth length in X_i, Q is the total number of lengths in X_i, and $Q <= H$. Then we obtain the largest length and use it as the length measurement l_i for the ith column:

$$l_i = \max_{j=1,\ldots,Q} c_j^i \quad (5.23)$$

In a similar way, the length measurements for all columns can be obtained:

$$L = \{l_1, l_2, \ldots, l_D\} \quad (5.24)$$

To represent the grass density, the length measurements are averaged over all columns, yielding the VOCGP for W:

$$VOCGP = a \times \frac{1}{D}\sum_{i=1}^{D} l_i \quad (5.25)$$

The estimated fuel load can be obtained based on the VOCGP:

$$\mathrm{F}_w = a \times VOCGP \quad (5.26)$$

where, a is a weight compensating for the different measurement units between VOCGP and the physically quantified fuel load in (5.1). Note that (5.23) and (5.26) correspond respectively to (5.5) and (5.6).

Algorithm 5.1 summarizes the procedure of calculating VOCGP for a sampling grass window. The VOCGP is obtained by two steps: (a) calculating the largest length of continuously connected grass pixels with the same dominant vertical orientation in each column and (b) taking the average value of all largest lengths for all columns. For the ith column in the window W, the algorithm starts by scanning the first pixel and checking whether the pixel $C_{i,j}$ is predicted as grass or non-grass using the ANN or CNN classifier, and then accounting the number *Length* of continuously connected grass pixels with a vertical orientation (i.e. $O_{x,y} = 2$) until a non-grass pixel or a non-vertical orientation pixel is found. Repeat the process for all pixels gives a series of lengths of connected grass pixels which are saved in C^i. However, only the largest length l_i is kept as it is the most important indicator of the grass height in the corresponding column. Small lengths are often generated by non-grass pixels, low grass pixels, or grass pixels without an absolutely dominant orientation. By contrast, high grass is expected to produce large lengths. The largest lengths of all columns can be obtained by repeating the above process. An 'average' operation is then performed on the largest lengths of all columns to obtain the VOCGP for W. We believe this 'average' operation is a necessary step to generate a reliable estimation of the average height of all grasses within each window by taking the grass density into account, as only when most columns have high lengths can we confirm in a high confidence that the window contains high and dense grasses and thus has high biomass.

Algorithm 5.1 : Pseudo-code of calculating VOCGP in a sampling grass window.

Input: A sampling window together with its pixel labels $A_{i,j}$ {grass; non-grass} and $O_{i,j}$ {vertical; non-vertical}.
Output: VOCGP.

```
SET L to empty
FOR each column in sampling window
    SET C to empty
    SET Lengt to zero
    FOR each row in sampling window
        IF A_{i,j} belongs to non-grass
            IF Lengt ≠ 0
                ADD Lengt to C
                SET Lengt to zero
            END IF
        ELSE
            IF O_{i,j} belongs to vertical
                INCREMENT Lengt
                IF O_{i,j} is last pixel in each column
                    ADD Lengt to C
                END IF
            ELSE
                IF Lengt ≠ 0
                    ADD Lengt to C
                    SET Lengt to zero
                END IF
        END IF
    END FOR
    FIND the longest length in C using (5.23)
    ADD the longest length to L
END FOR
GET VOCGP by averaging all elements in L using (5.25).
```

5.4 Experimental Results

This section evaluates the VOCGP approach for estimating grass biomass in three density categories of sparse, moderate and dense grasses using a sample image dataset. The accuracy is compared with the accuracy of human observation. The approach is further applied to the identification of fire-prone road regions using a test data of 100 video frames collected from roadside videos.

5.4.1 Roadside Data Collection

The data used for the experiments are images collected from a total of 61 sample roadside sites along the state roads within the Fitzroy region, Queensland, Australia, as displayed in Fig. 5.4. The sites were selected to cover various types and density levels of grasses, and the dominant species of grasses in those sites include Buffel, Blackspear, Dichanthium, Panicum, Rhodes, Stylosanthes, Thatch and Wire. The images of 1936 × 1296 pixels were taken using a Nikon D80 camera facing roadside grasses. To ensure consistency of the image size of sampling regions across sites, both the height of camera and the distance of the camera to the sampling regions were set to be approximately equal for all sites. During

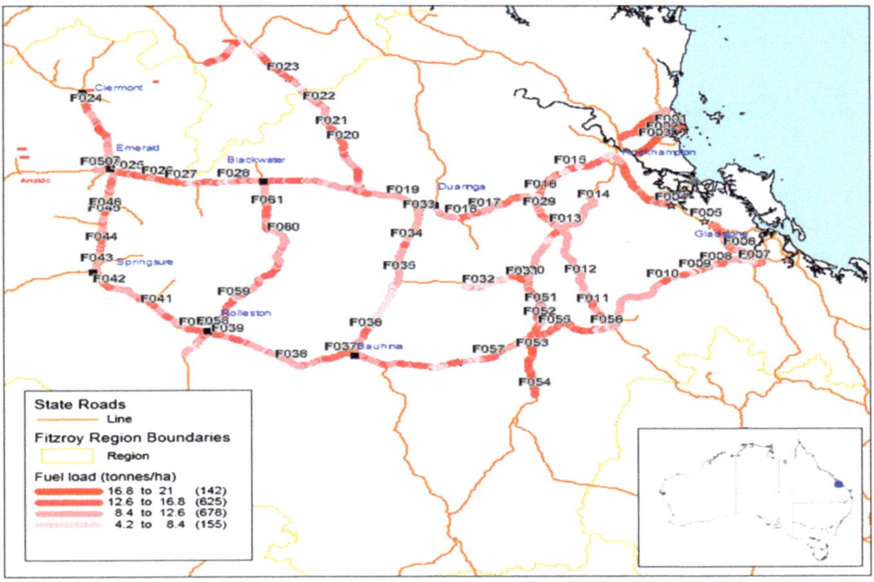

Fig. 5.4 Location distribution of 61 sample sites for taking images and collecting grass biomass data. These sites (F001 to F061) are chosen from the Fitzroy region, Queensland, Australia

Fig. 5.5 Image samples for sparse, moderate and dense grasses

experiments, the sampling grass windows were manually cropped from all images, which correspond exactly to sampling regions selected for 61 sample sites, and taken as inputs to the VOCGP approach.

Both subjective and objective measurements regarding the biomass and density level were then created respectively which serve as the ground truths for validating the accuracy of the VOCGP approach. To acquire subjective grass density, we classified all images into three categories of sparse, moderate and dense grasses based on human eye observation as samples shown in Fig. 5.5. The human classified categories of all images are shown in Table 5.1. To obtain the objective biomass, one square meter area was marked using a quadrat and the grass samples were cut, bagged, weighed and stored in the heater (700 °C) for drying up. The samples were taken out of the heater and weighed again after a few days (>72 h), and their fuel loads were estimated (tonnes/ha) using a dry weight and a standard formula. The estimated biomass for all images are shown in Table 5.2.

5.4.2 *Experimental Settings*

(1) Approach parameters. The ANN classifier has a structure of 23-16-2 neurons and is trained using a resilient backpropagation algorithm (goal error: 0.001 and maximum epochs: 200). The LeNet-5 CNN comprises seven layers and we use the implementation at https://github.com/sdemyanov/ConvNet. The training data

Table 5.1 Samples categorized into sparse, moderate, and dense grasses based on human observation

Density	Sample No
Sparse	F007, F008, F014, F017, F019, F022, F023, F024, F026, F030, F033, F034, F035, F038, F042, F043, F048, F050, F051, F056, F058, F060
Moderate	F002, F004, F006, F011, F013, F018, F020, F027, F029, F031, F032, F036, F039, F040, F041, F045, F047, F049, F052, F055, F061
Dense	F001, F005, F009, F010, F012, F015, F016, F021, F025, F028, F037, F044, F046, F053, F054, F057, F059

includes 650 grass and non-grass regions manually cropped from the video data collected by the DTMR, Queensland, Australia, covering both grass (green and brown grasses) and non-grass (road, tree, sky and soil) regions. The size of Gaussian filters is 7 × 7 pixels. The Gabor filters have four orientations $\theta = (0°, 45°, 90°, \text{ and } 135°)$, five scales $(\phi_m = \frac{f_{max}}{(\sqrt{2})^m}, m = 0, 1, \ldots, 4;$ $f_{max} = 0.25)$ and 11 × 11 Gabor kernels.

(2) Performance metrics: Two metrics are used: R^2 statistic and Root Mean Square Error (RMSE) between ground truth and estimated biomass. The RMSE is calculated using:

$$RMSE = \sqrt{\frac{1}{n}\sum_{t=1}^{n}\left(F_w^t - F_l^t\right)} \tag{5.27}$$

where F_w^t and F_l^t are estimated and ground truth biomass respectively for the tth sample image, and $n = 61$ is the number of images. The RMSE is an average error over five-fold cross validations, which group all images into five equal folds and in each validation, images in four folds are used for calculating the calibration factor by dividing the total biomass by the total VOCGP of all training samples, and images in the rest one fold for calculating the RMSE.

5.4.3 *Performance of Estimating Grass Biomass*

This part evaluates the performance of the VOCGP approach in estimating grass biomass for all samples. The VOCGP obtained from using an ANN or CNN classifier for grass region segmentation is defined as VocANN and VocCNN respectively.

Table 5.2 and Fig. 5.6 show the objective biomass and the corresponding VOCGP for all image samples, showing an overall consistency in the change tendency between their values although these values are calculated at different measurement units.

Figure 5.7 displays the correlation between them using a linear regression, indicating R^2 statistics 0.29 and 0.25 for VocANN and VocCNN respectively.

Since grass stems cannot grow perfectly in a vertical direction in real conditions, we evaluate the robustness of the VOCGP approach to non-vertical grass stems. Table 5.3 shows the RMSEs of VocANN obtained using images rotated by [−10, −5, 0, 5, 10] degrees. The RMSEs in original images are slightly lower than those

Table 5.2 Objective biomass and estimated VOCGP (i.e. VocANN and VocCNN for using ANN and CNN respectively) for sparse, moderate and dense grasses

Sparse				Moderate				Dense			
No	Bio	VocANN	VocCNN	No	Bio	VocANN	VocCNN	No	Bio	VocANN	VocCNN
F007	7.94	21.5	20.1	F002	8.31	31.0	23.6	F001	23.80	33.8	33.8
F008	6.10	32.0	28.6	F004	5.00	27.9	24.0	F005	20.57	40.2	35.0
F014	15.46	22.4	9.2	F006	10.68	22.8	20.6	F009	11.74	30.9	37.4
F017	4.28	14.7	14.0	F011	0.0	12.5	12.6	F010	16.01	46.6	23.9
F019	11.60	18.3	18.6	F013	11.93	29.5	22.3	F012	20.10	29.8	45.2
F022	6.78	19.8	19.1	F018	15.87	29.8	31.9	F015	32.10	60.8	25.8
F024	9.03	29.3	28.3	F020	14.74	29.8	25.8	F016	11.46	35.3	51.2
F026	4.20	26.6	21.6	F023	23.95	29.3	22.6	F021	11.95	32.9	32.0
F030	4.05	20.4	16.1	F027	7.12	27.5	40.3	F025	10.96	40.0	29.8
F033	11.75	13.0	13.4	F029	13.60	23.6	22.4	F028	21.24	19.5	36.5
F034	2.45	20.0	20.1	F032	13.50	28.9	32.9	F031	13.15	30.5	20.6
F035	4.15	24.2	22.1	F036	10.85	29.7	25.2	F037	16.00	48.9	31.8
F038	18.90	18.9	19.7	F040	14.85	29.0	26.7	F044	7.20	31.3	42.6
F039	10.90	18.9	17.6	F041	20.10	17.7	16.4	F046	14.85	30.3	22.3
F042	14.55	29.4	19.2	F045	5.50	29.0	32.8	F053	10.35	22.2	34.4
F043	3.45	22.4	21.2	F047	10.25	26.6	22.1	F054	22.85	33.0	21.3
F048	5.50	18.1	23.3	F049	11.45	29.6	27.3	F057	17.15	41.5	32.5
F050	6.85	15.8	13.4	F052	8.30	26.4	21.8	F059	12.20	24.9	39.2
F051	2.20	14.8	16.4	F055	13.10	29.5	21.7	–	–	–	–
F056	6.95	17.2	16.9	F061	8.15	25.6	20.5	–	–	–	–
F058	7.90	17.5	14.2	–	–	–	–	–	–	–	–
F060	4.15	18.2	17.1	–	–	–	–	–	–	–	–

Note Sample F003 is excluded due to blurred images captured for this site

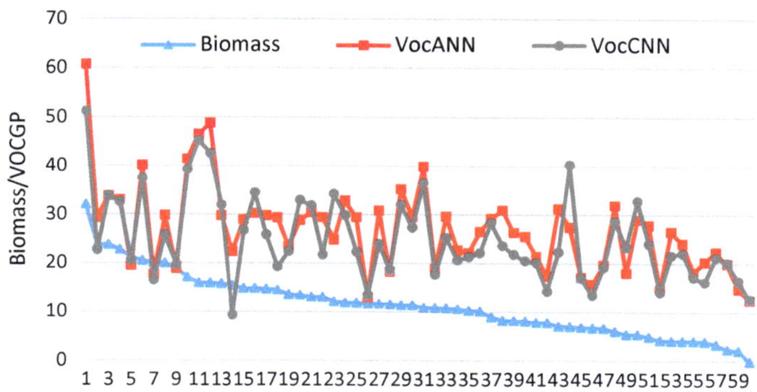

Fig. 5.6 Objective biomass and estimated VOCGP for all samples

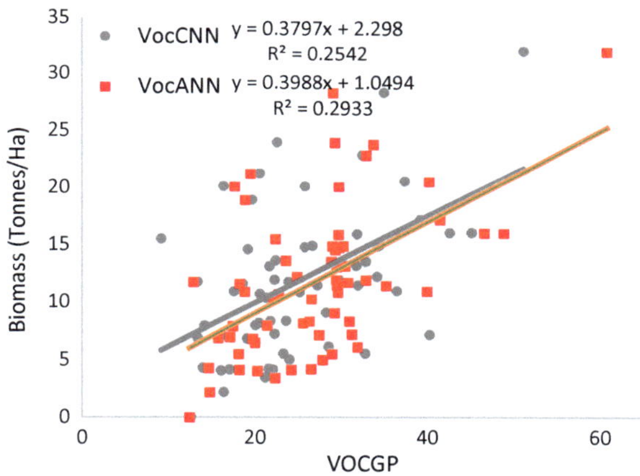

Fig. 5.7 Correlation between objective biomass and VOCGP for all samples

Table 5.3 Performance using images with different degrees of rotations

Rotation Degree	−10	−5	0	5	10
RMSE	5.95	5.95	5.84	5.87	6.02

in rotated images and tend to increase radically when the degree of rotation increases. However, the differences in RMSEs between the original and rotated images are relatively small, and this is largely due to the adoption of Gabor filter votes for orientation detection, which classifies grass stems with directions close to

Table 5.4 Performance comparisons of the VOCGP approach with human observation

	VocANN	VocCNN	Human observation
RMSE	5.84	5.66	5.49

a '90°' to a vertical orientation. The results confirm the robustness of the VOCGP approach to grass stems slightly deviated from a vertical direction.

We also compare the results of the VOCGP approach with human observation as shown in Table 5.4. The RMSE of human observation is calculated based on the mean biomass for sparse, moderate and dense grasses respectively. The VocANN and VocCNN show promising performance and have only 0.35 and 0.17 higher RMSE than human observation. The results indicate the potential of using machine learning technology for estimating grass biomass.

5.4.4 Performance of Predicting Grass Density

Figure 5.8 shows the mean values of the biomass and VOCGP over samples for sparse, moderate, and dense grasses. VocANN and VocCNN, and the mean biomass have similar positive correlations with the density of grasses, as a higher mean value of biomass (or VOCGP) is strongly associated with a higher density level of grasses. This is within our anticipation because both biomass and VOCGP depend on the height and density of grasses. The results demonstrate that VOCGP is able to accurately estimate the mean biomass within three density categories and predict the grass density.

To find lower/upper thresholds for classifying sparse, moderate and dense grasses, a box and whisker plot is drawn and shown in Fig. 5.9. The plot shows that, for biomass, VocANN and VocCNN, there are extreme minimum and maximum values for all three types of grasses and the majority of samples are concentrated around the median. Both VocANN and VocCNN tend to be more narrowly and unevenly distributed over the median than biomass, which might be a distinction between human observation classification and automatic machine

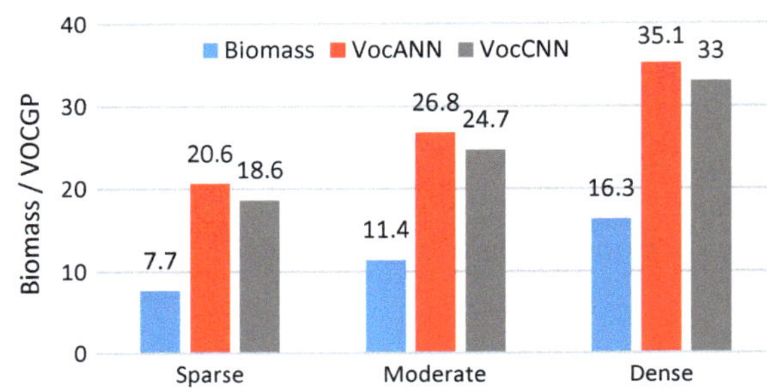

Fig. 5.8 Mean biomass and VOCGP for sparse, moderate and dense grasses

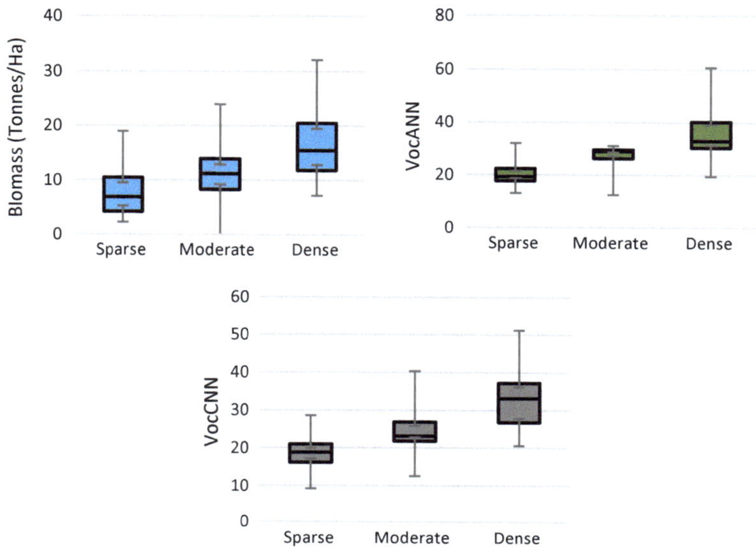

Fig. 5.9 Box and whisker plot of objective biomass and VOCGP for sparse, moderate and dense grasses

classification. The two thresholds of VocANN for classifying sparse and moderate, and dense grasses can be defined as 27 and 31 respectively.

Figure 5.10 shows best linear fits of objective biomass and estimated VOCGP of all samples to their corresponding categories of sparse, moderate and dense grasses. R^2 statistics of 0.30, 0.47 and 0.45 have been observed for biomass, VocANN and VocCNN respectively with respect to their correlations with density categories, indicating the consistency between objective biomass measurement and the VOCGP approach in the prediction of grass density. Both VocANN and VocCNN have higher correlations than objective biomass, which confirms the effectiveness of the VOCGP approach in predicting grass density categories.

5.4.5 Fire-Prone Region Identification

To prove the effectiveness of the VOCGP approach in identifying fire-prone road regions with high biomass, we conduct experiments on the roadside video data for a state road No. 16A in the Fitzroy region collected by the DTMR, Queensland, Australia. We select 100 frames from a total of 22 videos so that these frames are equally distributed over the whole road with at least 200 m apart between frames. In each frame, 15 overlapped sample windows as shown in Fig. 5.11 are selected and manually annotated into sparse, moderate or dense grasses.

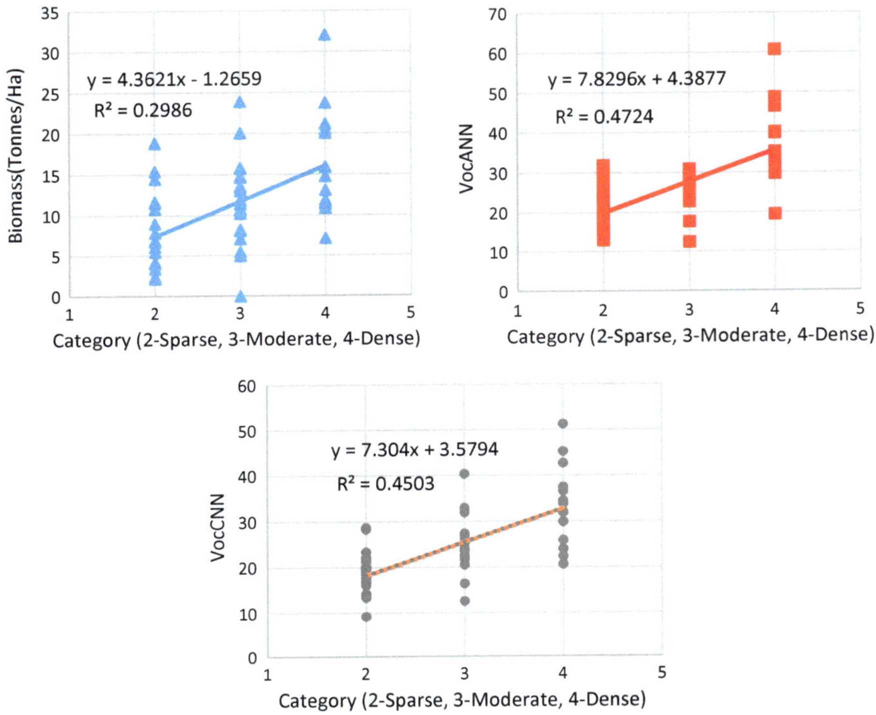

Fig. 5.10 Best linear fit of objective biomass and VOCGP to three grass categories

We firstly apply the two thresholds for VocANN (27 and 31) obtained from Fig. 5.9 to classify the density category in all windows. The overall accuracies are 81.7, 75.2 and 61.2% for sparse, moderate and dense grasses respectively. Table 5.5 shows the confusion matrix of classifying all windows into sparse, moderate or dense grasses using the estimated VocANN. The overall accuracy is 73.2% for three categories. Sparse grasses are the easiest for correct classification with accuracy of 81.7%, while dense grasses are the most difficult with accuracy of 61.2% and a significant portion (27.7%) of dense windows are misclassified as moderate. This is largely due to the confusion between dense and moderate windows, for which even humans may find hard to classify. For instance, there are a large number of windows that are covered by both high and low grasses. For some types of grasses, they have high heights and high VocANN, but actually with low biomass. Figure 5.11 shows ground truths, grass segmentation results, and estimated VocANN in sample images.

We then take the average VocANN over all 15 windows as an indicator of the biomass yield in each image. For an image with dense and high grasses, it is expected that most windows have high VocANN, leading to a high average value and a high chance of fire hazards. Figure 5.12 displays the average VocANN for all

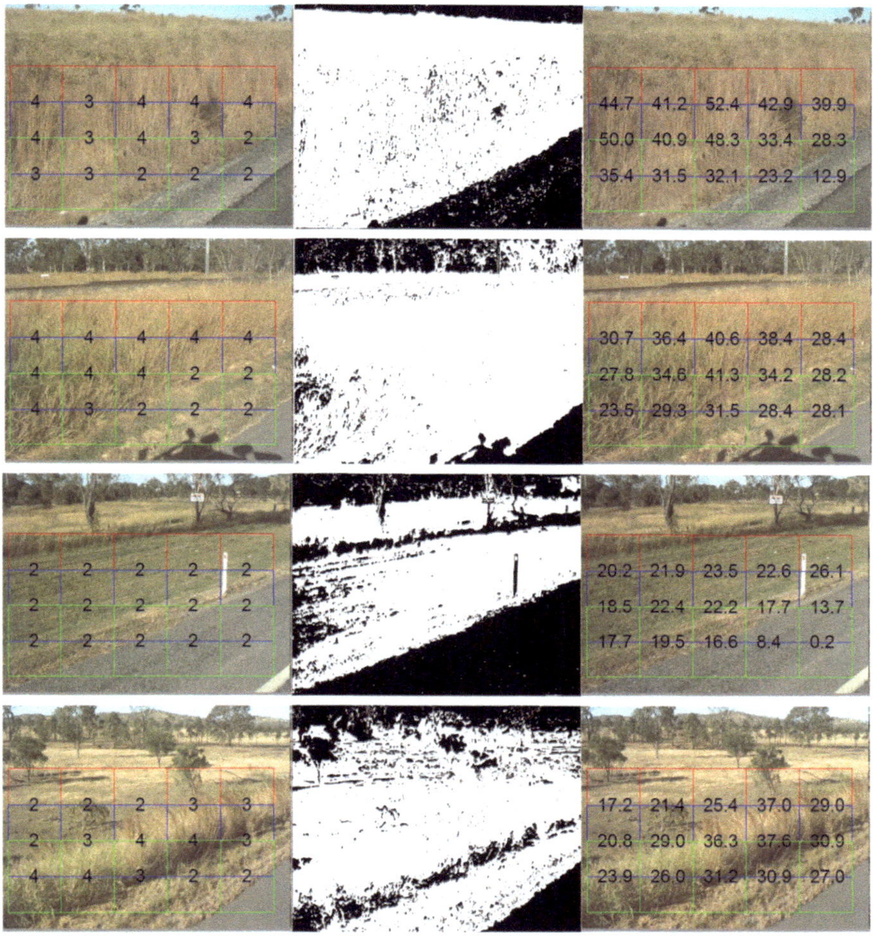

(a) Window Ground Truth (b) Grass Segmentation (c) Estimated VOCGP

Fig. 5.11 Distribution and classified results of 15 sampling grass windows in image frames. **a** Ground truth: 2-sparse, 3-moderate, and 4-dense; **b** grass segmentation results: *white* grass pixels and *black* non-grass pixels; **c** estimated VOCGP, i.e. VocANN

Table 5.5 Confusion matrix of sparse, moderate and dense grasses using estimated VocANN

	Sparse	Moderate	Dense
Sparse	81.7	13.5	4.8
Moderate	10.0	75.2	14.8
Dense	11.1	27.7	61.2

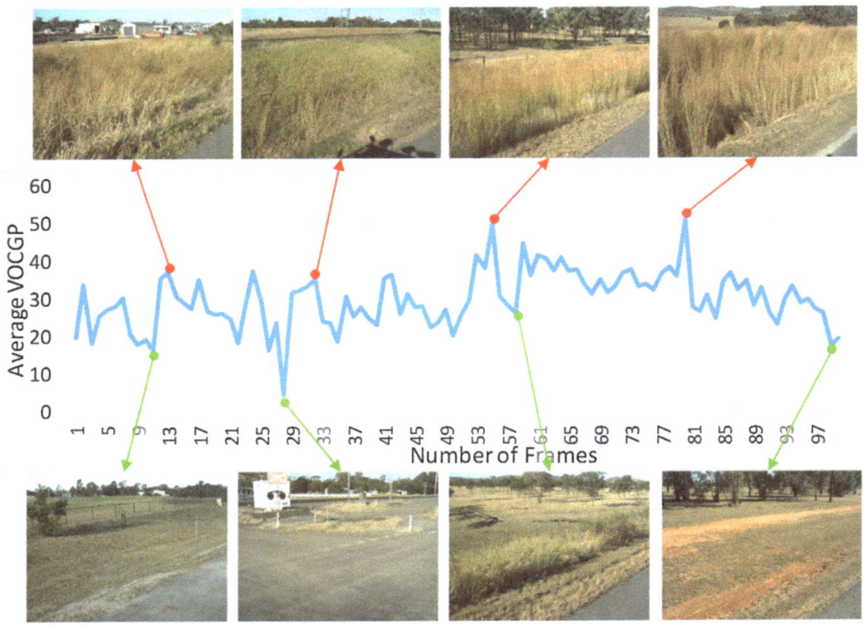

Fig. 5.12 Illustration of using the average VocANN over 15 sampling windows in each image for fire-prone region identification. The frames are ordered according to their locations on the state road No. 16A in the Fitzroy region. The locally highest and lowest VocANN are matched accurately with the corresponding grass density levels

100 frames according to their location on the road. Eight typical frames with locally high, low or middle VocANN are also displayed and it can be seen that the approach using the average VocANN produces quite promising results and the locally highest and lowest VocANN are matched accurately with the grass density levels in their corresponding frames. The locations of frames with the locally highest average VocANN can be identified as fire-risk regions.

5.5 Discussions

The main lessons learnt from the experimental results are as follows.

(1) Our experimental results confirm the feasibility of using machine learning techniques to automatically estimate the biomass of grasses. It was found that the predicted VOCGP has a similar overall tendency with the objective biomass for all samples, with RMSEs close to human observation.

(2) Both the mean biomass and predicted VOCGP have a similar positive correlation with the density of grasses, i.e. sparse, moderate, and dense grasses. The

approaches of VocANN and VocCNN have R^2 equal to 0.47 and 0.45 respectively in the prediction of grass density, and 0.29 and 0.25 respectively in the estimation of biomass. The results confirm the effectiveness of the VOCGP approach in predicting grass density categories and biomass.

(3) Evaluations on a set of manually annotated sampling windows show an average accuracy of 73.2% for classifying sparse, moderate and dense grasses, and the lowest accuracy of 61.2% is obtained for moderate grasses, largely due to the confusion in windows containing both high and low grasses. We further confirm the feasibility of using the average VocANN over all windows in each image to identify fire-prone regions on a state road in the Fitzroy region.

The estimation of grass biomass using machine learning techniques needs to consider several impacting factors:

(1) Parameters of the sampling window, including the location, size, shape, etc. The parameters of the window in 2D static images may not correspond exactly to those of real-world 3D sampling grass regions in field surveys, which may lead to biased estimation results. While the location is relatively easy to be determined, the size of the window should be set properly based on the resolution of the image and the range of grass regions in the image, as a small window is unable to cover the whole plant height, while a large one is likely to include unpredicted objects. Strictly speaking, using a fixed size of windows is only workable for images with the same resolution and a similar height of grass regions.
(2) Accuracy of grass segmentation algorithms. In the case that non-grass pixels are misclassified into grass pixels, a VOCGP higher than its actual value will be got, resulting in a false prediction decision of high biomass, and vice versa. In addition, the VOCGP calculation is also sensitive to small isolated non-grass pixels, which break the connectivity of grass pixels and lead to a falsely low VOCGP. A solution might be introducing a post-processing step to remove isolated non-grass pixels, such as morphological opening or regional smoothing over a neighbourhood (e.g. superpixels). It should be noted that accurate grass segmentation in complex scenes is still a challenging topic itself.
(3) Methods of detecting the dominant local orientation at each pixel. The parameters of Gabor filters such as the size of kernels and the number of scales might impact the detection accuracy, and other detection methods such as pixel intensity and edge detectors are also worth investigating for more effective detection.

5.6 Summary

The estimation of roadside grass biomass using machine learning techniques is still a largely unexplored area. A VOCGP approach is described in this section for the estimation of the biomass of roadside grasses based on the grass height and density. The evaluations are based on a sample image dataset from 61 sites along the state roads within the Fitzroy region, Queensland, Australia with ground truths of objective biomass and subjective categories of sparse, moderate and dense grasses. We compare non-deep and deep learning algorithms for grass segmentation in the VOCGP approach—VocANN and VocCNN, and show that they have R^2 equal to 0.47 and 0.45 respectively in the prediction of grass density, 0.29 and 0.25 respectively in the estimation of biomass, and RMSEs close to human observation. There are no big differences in the prediction results between using non-deep and deep learning techniques. The VOCGP approach demonstrates promising results on automatically identifying fire-prone regions on a roadside video data. One possible future work is to investigate probability-based soft decisions on both grass region segmentation and dominant orientation calculation rather than binary decisions.

References

1. Y.F. Vazirabad, M.O. Karslioglu, LIDAR for biomass estimation, in *Biomass—Detection, Production and Usage* (INTECH Open Access Publisher, 2011)
2. H.W. Zub, S. Arnoult, M. Brancourt-Hulmel, Key traits for biomass production identified in different Miscanthus species at two harvest dates. Biomass Bioenerg. **35**, 637–651 (2011)
3. D. Ehlert, R. Adamek, H.-J. Horn, Laser rangefinder-based measuring of crop biomass under field conditions. Precis. Agric. **10**, 395–408 (2009)
4. N. Tilly, D. Hoffmeister, Q. Cao, V. Lenz-Wiedemann, Y. Miao et al., Transferability of models for estimating paddy rice biomass from spatial plant height data. Agriculture **5**, 538–560 (2015)
5. N. Tilly, H. Aasen, G. Bareth, Fusion of plant height and vegetation indices for the estimation of barley biomass. Remote Sens. **7**, 11449–11480 (2015)
6. C. Royo, D. Villegas, Field measurements of canopy spectra for biomass assessment of small-grain cereals, in *Biomass—Detection, Production and Usage* (INTECH Open Access Publisher, 2011)
7. T. Ahamed, L. Tian, Y. Zhang, K.C. Ting, A review of remote sensing methods for biomass feedstock production. Biomass Bioenerg. **35**, 2455–2469 (2011)
8. T. Sritarapipat, P. Rakwatin, T. Kasetkasem, Automatic rice crop height measurement using a field server and digital image processing. Sensors **14**, 900–926 (2014)
9. Z. Juan, H. Xin-yuan, Measuring method of tree height based on digital image processing technology, in *1st International Conference on Information Science and Engineering (ICISE)*, 2009, pp. 1327–1331
10. H. Dianyuan, W. Chengduan, Tree height measurement based on image processing embedded in smart mobile phone, in *International Conference on Multimedia Technology (ICMT)*, 2011, pp. 3293–3296
11. H. Dianyuan, Tree height measurement based on image processing with 3-points correction, in *International Conference on Computer Science and Network Technology (ICCSNT)*, 2011, pp. 2281–2284

12. N. Soontranon, P. Srestasathiern, P. Rakwatin, Rice growing stage monitoring in small-scale region using ExG vegetation index, in *11th International Conference on Electrical Engineering/Electronics, Computer, Telecommunications and Information Technology (ECTICON)*, 2014, pp. 1–5
13. J. Payero, C. Neale, J. Wright, Comparison of eleven vegetation indices for estimating plant height of alfalfa and grass. Appl. Eng. Agric. **20**, 385–393 (2004)
14. J. Llorens, E. Gil, J. Llop, A. Escolà, Ultrasonic and LIDAR sensors for electronic canopy characterization in vineyards: advances to improve pesticide application methods. Sensors **11**, 2177–2194 (2011)
15. B. St-Onge, Y. Hu, C. Vega, Mapping the height and above-ground biomass of a mixed forest using LIDAR and stereo IKONOS images. Int. J. Remote Sens. **29**, 1277–1294 (2008)
16. G. Grenzdörffer, Crop height determination with UAS point clouds. Int. Arch. Photogramm. Remote Sens. Spat. Inf. Sci. **1**, 135–140 (2014)
17. K. Yamamoto, T. Takahashi, Y. Miyachi, N. Kondo, S. Morita et al., Estimation of mean tree height using small-footprint airborne LIDAR without a digital terrain model. J. For. Res. **16**, 425–431 (2011)
18. J. Cai, R. Walker, Height estimation from monocular image sequences using dynamic programming with explicit occlusions. IET Comput. Vis. 16, 149–161
19. L. Zhang, T.E. Grift, A LIDAR-based crop height measurement system for *Miscanthus giganteus*. Comput. Electron. Agric. **85**, 70–76 (2012)
20. C. Rasmussen, Grouping dominant orientations for ill-structured road following, in *IEEE Conference on Computer Vision and Pattern Recognition (CVPR)*, 2004, pp. 470–477
21. W.T. Freeman, E.H. Adelson, The design and use of steerable filters. IEEE Trans. Pattern Anal. Mach. Intell. **13**, 891–906 (1991)
22. F. XiaoGuang, P. Milanfar, Multiscale principal components analysis for image local orientation estimation, in *Conference Record of the Thirty-Sixth Asilomar Conference on Signals, Systems and Computers*, 2002, pp. 478–482
23. A.K.P. Meyer, E.A. Ehimen, J.B. Holm-Nielsen, Bioenergy production from roadside grass: a case study of the feasibility of using roadside grass for biogas production in denmark. Resour. Conserv. Recycl. **93**, 124–133 (2014)
24. J. Shotton, M. Johnson, R. Cipolla, Semantic texton forests for image categorization and segmentation, in *IEEE Conference on Computer Vision and Pattern Recognition (CVPR)*, 2008, pp. 1–8
25. J. Winn, A. Criminisi, T. Minka, Object categorization by learned universal visual dictionary, in *Tenth IEEE International Conference on Computer Vision (ICCV)*, 2005, pp. 1800–1807
26. J. Shotton, J. Winn, C. Rother, A. Criminisi, Textonboost for image understanding: multi-class object recognition and segmentation by jointly modeling texture, layout, and context. Int. J. Comput. Vis. **81**, 2–23 (2009)
27. Y. Lecun, L. Bottou, Y. Bengio, P. Haffner, Gradient-based learning applied to document recognition, in *Proceedings of the IEEE*, vol. 86, (1998), pp. 2278–2324
28. J.K. Kamarainen, V. Kyrki, H. Kalviainen, Invariance properties of gabor filter-based features-overview and applications. IEEE Trans. Image Process. **15**, 1088–1099 (2006)

Chapter 6
Conclusion and Future Insight

In this chapter, we present several recommendations for future research work based on the experimental results obtained using various non-deep and deep learning techniques. We also highlight challenges in this field and discuss new opportunities and applications.

6.1 Recommendations

Based on experimental results from various deep and non-deep learning techniques on roadside data analysis, we can make following recommendations:

(1) Discriminative feature extraction. It is advisable to consider color, texture and contextual information for more robust segmentation of roadside objects. The experimental results using both deep and non-deep learning techniques show that a combination of color and texture produces higher classification accuracy than using them alone, for most roadside objects. The incorporation of local and global contextual information, such as the CAV features, yields significant improvements to the performance compared with using visual features alone, including color and texture. The superior performance of CAV features is largely due to their advantages of capturing both long- and short-range label dependencies between objects, being able to adapt to the image content, and preserving both relative and absolute location information.

(2) Contextual information using deep leaning techniques. Compared to existing graphical models proposed for encoding both local and global contextual information for object categorization, deep learning techniques have advantages of automatically encoding contextual information, extracting visual features, and integrating both of them inherently in the deep learning architecture. We introduced a deep learning network for object segmentation and confirmed its state-of-the-art performance in real-world benchmark datasets.

B. Verma et al., *Roadside Video Data Analysis: Deep Learning*,
Studies in Computational Intelligence 711, DOI 10.1007/978-981-10-4539-4_6

(3) Existing patch based feature extraction techniques suffer from the boundary problem, which means that extracting features in boundaries between objects can unavoidably introduce noise into the feature set, due to a fixed rectangular shape of patches. To handle noise in regional boundaries, PPS features are presented based on segmented superpixels, and they demonstrate higher accuracy for object segmentation than both pixel based and patch based features on natural roadside data. Thus, it is still necessary to investigate more effective techniques that can overcome the boundary problem to further improve the performance of object segmentation.

(4) Enforcing constraints on spatial locations of objects is beneficial to achieving higher accuracy in detecting and segmenting roadside objects. For instance, sky is likely to be at the top part of roadside images. However, the use of spatial constraint is largely limited to pre-knowledge about the situations of a specific application. Thus, it is advisable to carefully pre-analyze the context of a specific application in designing suitable techniques and enforcing appropriate constraints accordingly for robust results.

(5) Rather than solely depending on the training data, properly utilizing the local characteristics in the test image can help create robust algorithms capable of automatically adapting locally to the content of the test image, leading to more robust results in handling realistic challenges, such as noise and lighting variations. As an instance, the superpixel merging approach—SCSM, which was designed to consider both general characteristics of all objects on the training data and local properties in the test image, has shown good performance in removing misclassification error in the results of traditional classifiers such as ANN.

(6) CNN with automatic feature extraction performs well in computer vision tasks and is—applicable to high noise and broad variations that typify real-world applications. The CNN performs well but is not necessarily the best choice for small image classification tasks compared with a traditional MLP. Therefore, it is recommended to consider using CNN for datasets with a reasonably large amount of data, while for small datasets, both CNN and traditional non-deep learning classifiers may need to be comparatively considered. However, the results become different when an assemble strategy is adopted, where an ensemble of CNNs performs better than using a single MLP or CNN classifier, and an assemble of MLP classifiers. Therefore, it is advisable to consider an ensemble of CNNs for handling object classification problems in real-world datasets.

(7) The majority voting method for combining multiple classifiers to reach a final decision for object classification achieves higher accuracy for dense versus sparse grass discrimination than using individual classifiers, although the differences in the accuracy are not statistically significant based on an ANOVA test. The results indicate that even a simple combination strategy is able to result in more accurate performance for object classification and, therefore, it is generally recommended to consider the adoption of multiple classifiers in

handling variations in the appearance of objects and the environments in real-world roadside datasets.

(8) We present an ensemble learning approach which generates and fuses different versions of neural networks for road object detection. The approach uses different seeding points to partition data into multiple layers and generates clusters at each layer. A neural network is generated for each cluster and each layer. A majority vote over those neural networks shows big improvements in the classification rate over SVM, hierarchical, and clustering approaches for road object detection. Thus, creating different versions of an individual classifier by tailoring each of them to a different pattern in the data helps to increase the diversity of the ensemble classifier and eventually leads to a higher performance. As a recommendation, individual classifiers may not need to be restricted to the same type and they can be different types so that the diversity can be further improved.

(9) We conduct a case study to confirm the feasibility of using machine learning techniques to automatically estimate the biomass and density of grasses. The predicted VOCGP has a similar overall tendency with objective biomass for all samples, with RMSE close to human observation. Although grass stems can grow into different directions, the connectivity of grass pixels in a vertical direction seems to be a reliable indicator of the height of grasses. Compared with using ANN for grass region segmentation, using CNN has a slightly lower correlation with objective biomass, but a lower RMSE. This indicates that CNN and ANN have performed similarly in this case study.

6.2 New Challenges

(1) For biomass estimation in digital images, one challenge is how to determine the sampling image region, including the location, size, shape, etc. The parameters of the region in 2D images may not correspond exactly to those of the 3D sampling grass region in field surveys, which leads to biased estimation results. Although manual annotation of sampling regions is possible, it is often time-consuming and may significantly restrict the automation of the whole system. Another challenge is how to make estimated and objective biomass directly comparable, as they are not measured by the same unit, i.e. pixel versus tonnes/ha. One solution is to set a scale factor between the two values as done in the case study, but another question arises regarding how to calculate the factor, which may significantly impact the estimation accuracy.

(2) It is still a challenge to accurately measure the distance of roadside grasses to the road boundary in 2D images. The distance is traditionally considered as a key factor that has big impact on the fire risk level of roadside grasses, as grasses close to the road normally present bigger risk while those far away from

the road have less risk. Due to the loss of depth information, far and near grasses can overlap at the same region in 2D images, which makes it difficult to measure precisely their distances to the road separately. One possible solution is to consider the collection of 3D data about roadside objects, but this increases the difficulty of data collection and puts higher requirements on the equipment.

(3) The calculation of the parameters of roadside objects is sensitive to the distance between the object and camera. When the distance changes, the size of the objects in the captured images varies accordingly. This generates a big challenge for measuring the actual values of parameters of roadside objects, such as height and area, because the actual measurement unit per pixel may change substantially in different images. For instance, a zoom-in region of grasses leads to a higher estimated height than a zoom-out region of the same grasses. Therefore, many present approaches simply assume a fixed distance between objects and the camera.

(4) The current literature still lacks a comprehensive public roadside dataset with well annotation of pixel-wise ground truths of objects. Creating such a dataset faces challenges such as data copyright issues, privacy issues, and the requirement of labor, time, and effort. The generation of a specific dataset may not able to fit the requirements for other evaluation purposes in terms of the number and type of object categories, the size of the data, the resolution and change rate of frames, the environmental condition, the location, etc.

6.3 New Opportunities and Applications

(1) One possible future application is the development of a mobile system for fire-prone road region identification, which allows roadside residents or drivers to take photos of roadside grasses using their mobile devices and provides a prediction of the fire risk level of the grasses. The residents can then report the locations of the sites with high fire risk to relevant government authorities, so that service people can be sent to those locations to take necessary measurements to remove the risk. This may greatly facilitate relevant authorities in detecting, monitoring, and maintaining safe road conditions in a more efficient and effective manner.

(2) Investigating techniques that are able to detect and recognize roadside objects such as road signs can significantly boost the development and deployment of smart vehicles, which can provide critical road information or timely alerts to drivers and improve the safety of driving, particularly in severe weather conditions or at dangerous road sites. The drivers can be pre-informed of both roadside and road conditions ahead to adjust the driving behaviors accordingly, such as petrol stations and rest areas.

(3) Analysis of roadside data content is also helpful in promoting effective maintenance of road conditions, such as detecting the situations where the road boundary lines became invisible, road surface was damaged, or road fence was broken. Without implementing a procedure of sending staff to visually check those road sites, the transport authority can carry out proper maintenance at those identified sites, ensuring timely maintenance actions. As a supplementary application, the analysis can also be used to automatically detect roadside advertising such as billboard signs and business signs, construction activities, and equipment installation where a vendor have not got legal permits, violate legal requirements, or have potential impacts on the safety and efficiency of road users.

(4) Considering the increasing popularity and robustness to environmental variations of invisible features in large field tests, it is advisable to consider a combination of visible and invisible features for more robust and accurate analysis in real-world environments. Visible features are better in representing visual appearance and structure of objects in good conditions, while invisible features are more robust against environmental challenges. An object segmentation system capable of intelligently switching between them is expected to work more robustly in real-world situations.

MIX
Papier aus verantwortungsvollen Quellen
Paper from responsible sources
FSC® C105338

If you have any concerns about our products,
you can contact us on
ProductSafety@springernature.com

In case Publisher is established outside the EU,
the EU authorized representative is:
Springer Nature Customer Service Center GmbH
Europaplatz 3, 69115 Heidelberg, Germany

Printed by Libri Plureos GmbH
in Hamburg, Germany